THE HUNT FOR LIFE ON MARS

BY THE SAME AUTHOR

THE HUNT FOR LIFE ON MARS

DONALD GOLDSMITH

A DUTTON BOOK

DUTTON
Published by the Penguin Group
Penguin Books USA Inc., 375 Hudson Street, New York, New York 10014, U.S.A.
Penguin Books Ltd, 27 Wrights Lane, London W8 5TZ, England
Penguin Books Australia Ltd, Ringwood, Victoria, Australia
Penguin Books Canada Ltd, 10 Alcorn Avenue,
Toronto, Ontario, Canada M4V 3B2
Penguin Books (N.Z.) Ltd, 182–190 Wairau Road, Auckland 10, New Zealand

Penguin Books Ltd, Registered Offices:
Harmondsworth, Middlesex, England

First published by Dutton, an imprint of Dutton Signet,
a division of Penguin Books USA Inc.
Distributed in Canada by McClelland & Stewart Inc.

First Printing, February, 1997
10   9   8   7   6   5   4   3   2   1

CIP data is available.

Printed in the United States of America
Set in Bitstream Carmina
Designed by Stanley S. Drate/Folio Graphics Co. Inc.

This book is printed on acid-free paper. ⊗

To my daughter Rachel—
on a quest of her own

Acknowledgments

I would like to acknowledge my good fortune in having received assistance from many friends and scientists (not mutually exclusive categories) who have helped me to understand and to explain the different aspects of the search for life on Mars. Without their aid, I could not have completed this book.

I am grateful to all those who so kindly provided comments, explanations, suggestions, and advice for this work: Edward Anders, Bruce Armbruster, Richard Ash, Peter Baker, Justine Barletta, John Baross, John Barrow, Donald Bogard, Kenneth Brecher, Margaret Brose, Michael Carr, William Cassidy, Simon Clemett, Jeff Cuzzi, David Deamer, David Des Marais, Russell Doolittle, George Field, Richard Frankel, Imre Friedmann, Owen Gingerich, Larry Gold, Paul Goldsmith, Rachel Goldsmith, Hyman Hartman, Ralph Harvey, Jerry Heymann, Norman Horowitz, James Kasting, Dennis Kent, Joseph Kirschvink, Andrew Knoll, Geoff Marcy, Lynn Margulis, Lawrence Marschall, Ursula Marvin, Christopher McKay, David McKay, David Mittlefehldt, David Morrison, Doug Nash, David Nyquist, Leslie Orgel, Tobias Owen, Norman Pace, Christopher Romanek, Anneila Sargent, J. William Schopf, Roberta Score, Frank Shu, Lynn Simarski, George Smoot, Michael Soule, Larry Squire, Clifford Stoll, Jack Szostak, Brian Toon, Allen Treiman, Carol Trussel, Hojatollah Vali, David Vassar, Meenakshi Wadhwa, and Richard Zare.

"This rock from Mars

will change the way

we think about

life in the

universe."

CONTENTS

Contents

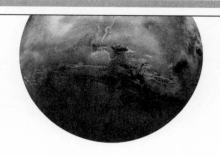

"Only seven people

in the world

know this.

NASA has found

life on Mars!"

PREFACE

Washington, D.C., July 1996: In an exclusive, expensive hotel a few blocks from the White House, the President's chief political advisor has arrived at his weekly tryst with a high-powered "escort." Seeking to amplify the aphrodisiac aura bestowed upon him by his access to power, and blissfully unaware that his paramour has arranged to sell him out to a weekly tabloid, he whispers secrets from the inner circles of government into her shell-like ear. "Only seven people in the world know this," he intones breathlessly. "NASA has found life on Mars!"

Although the dialogue may be fanciful, the passage of information described above apparently did occur. Not many specific discoveries rise to the level of interest attained by the news of possible Martian life. "It exceeded my wildest dreams to think that my research could be part of a White House–scandal pillow talk," said Richard Zare, one of the key players on the team that found signs of ancient life in a rock from the planet Mars. Even though he is one of the most important scientists in the United States, Zare was quite content to pursue his research in obscurity, so far as the general public was concerned. The sudden, totally unexpected announcement of possible life on Mars changed Zare's life, at least for a few months. The news has also affected the attitudes that many scientists, as well as many members of the wider public, have about humanity's place in the universe.

Preface

As a scientist turned science popularizer, I aim to help these changes occur, and to take the reader of this book on a visit to a "scientific courtroom," in which scientists will present the data pointing toward life on Mars and will debate the conclusions that follow from this evidence. The scientific detectives will help us to judge the reliability of the testimony contained within the Martian rock, and even more to assess the implications of that evidence. Like a complex murder investigation, this case involves a large set of scientific fields of research: astronomy, biology, chemistry, and geology, for starters, plus more exotic, barely born areas such as "exobiology" (life in places other than Earth) and "origin-of-life studies," a field that remains in relative infancy, unable to produce laboratory results that duplicate what happened on our planet some four billion years ago. The scientific testimony will range from essentially undisputed items, such as the great age of the rock that fell from Mars, to fully and eagerly disputed ones, such as the interpretation of the microscopic "ovoids" found within this meteorite.

Like the detectives at a crime scene, scientists involved in the hunt for life on Mars must search for clues while they are not even sure what form those clues may take, and therefore must remain ever alert for the Martian analogue to a murder weapon or blood-stained glove that a criminal might drop in his haste. Like the traces that criminalists may gather and analyze, the evidence for ancient life on Mars will persuade different people to different degrees. Just as society leaves the final criminal judgment not to experts but to a (so we hope) representative jury of citizens, each of us must decide, if we care to, the extent to which the case for ancient life

Preface

on Mars seems compelling. As part of that judgment, we must, like a criminal jury, decide how much weight to give to the testimony of various experts, each with specialized areas of knowledge bearing on a different phase of the case, sometimes contradicting one another, rarely in absolute agreement. As we explore the possibility of life on Mars, I shall try to show you how to appreciate the strands of scientific endeavor that have woven what is unquestionably a marvelous story, the tale of an extraordinary rock that has secrets to tell us about the history of life on two planets.

"Extraordinary

claims

require

extraordinary

evidence."

THE DAY
THE WORLD
CHANGED

On August 7, 1996, President Bill Clinton, standing outside the White House in Washington, D.C., announced that a meteorite found in Antarctica contains evidence suggesting the existence of ancient life on Mars. The meteorite had in fact come from Mars; the abundance ratios of elements found in its interior match the Martian values. Even more surprising, age-dating techniques based on the decay of radioactive elements had set its age at four and a half billion years, making this Martian meteorite older than *any* other rock ever found on Earth or the moon. Still more extraordinary was the fact that this ancient four-pound visitor from Mars contained several types of evidence implying that incredibly tiny, long-dead forms of life had once inhabited crevices in the rock.

The Hunt for Life on Mars

"Like all discoveries," the well-briefed President stated, "this one will and should continue to be reviewed, examined, and scrutinized. . . . I am determined that the American space program will put its full intellectual power and technological prowess behind the search for further evidence of life on Mars." Daniel Goldin, the administrator of the National Aeronautics and Space Administration (NASA), had more specific comments for the press. "I want everyone to understand that we are not talking about 'little green men,'" he emphasized. "These are extremely small, single-cell structures that somewhat resemble bacteria on Earth. There is no evidence or suggestion that any higher life-form ever existed on Mars."

The President's announcement summarized what the scientists involved in the discovery had just stated at a press conference at NASA Headquarters—a conference that had been arranged in great haste after the news about life on Mars had somehow leaked into the public arena. Scientific protocol requires that no announcement of any new discovery should occur until the results have been published in a peer-reviewed journal. Until the official date of publication of the Mars story, which in this case was August 16, the researchers who wrote the article, the scientists who reviewed it, and the staff of the publication in which it was to appear all had to maintain silence. This type of temporary embargo on the news aims to prevent speculation about results only imperfectly described, and to assure that proper credit will accrue to those who have advanced scientific knowledge.

For the Mars meteorite, as for many other discoveries, the relevant journal was *Science*, published by the

The Day the World Changed

American Association for the Advancement of Science, the most widely read and highly respected scientific journal in the United States, and one whose embargoes of news announcements have usually succeeded. But early in August, something had gone wrong with the protocol: News of the discovery of ancient life on Mars had begun to circulate widely, and a rumor was loose that the full text of the *Science* article had appeared on the World Wide Web. Just who leaked the story from Mars, and why, remains shrouded in the mists of recent history. Since the leaks were irretrievable, however, NASA felt that it must seize the moment and allow the news media access to the scientists who were about to publish their discoveries.

Thus, on August 7, 1996, with a special dispensation from *Science* magazine, five men and one woman, accompanied by Daniel Goldin and NASA's associate administrator for space science, Wesley Huntress Jr., appeared before the press to tell the story of the rock from Mars and its implications. Two of these scientists, David McKay and Everett Gibson, are geologists who work at NASA's Johnson Space Center (JSC) in Houston; another, Kathie Thomas-Keprta, is an expert in electron microscopy who also works at the Johnson Space Center, but as a contract employee of the Lockheed Martin Corporation; the fourth, Hojatollah Vali, is a geologist at McGill University in Toronto; the fifth, Richard Zare, is an analytical chemist at Stanford University; while the sixth, J. William Schopf from UCLA, is an expert on the earliest evidence for life on Earth.

Schopf was not actually a member of the team that had analyzed the rock from Mars. NASA had wisely invited him to demonstrate the principle of organized skepticism that allows science to succeed by subjecting all assertions, especially extreme ones, to doubting scrutiny. Before the conference was over, Schopf would twice quote a line he attributed to Carl Sagan: "Extraordinary claims require extraordinary evidence."

Every scientist and journalist in NASA's auditorium remembered the brief flash of news several years earlier about cold fusion, the "greatest discovery since fire," when preliminary results of an experiment had been untimely rushed before the public eye. In March 1989, two chemists named Stanley Pons and Martin Fleischmann claimed to have seen the results of nuclear fusion at room temperatures, rather than at the multimillion-degree temperatures of the regions where nuclear fusion occurs inside the sun and other stars, and in hydrogen bombs as well. For a few weeks, cold fusion appeared to offer an answer to the prayers of an energy-starved world, and those skeptics who doubted the reality of the phenomenon seemed no more than disappointed would-be inventors. Before long, the University of Utah, where Pons chaired the chemistry department, retained a high-powered business consultant named Ira Magaziner to help it secure federal funding for cold-fusion research. Magaziner, who would later head a high-level White House task force on health reform, darkly reminded Congress that "it is midnight in Japan. Thousands of Japanese scientists are trying to . . . develop commercial applications . . . of this new science."

The new science turned out to be nonscience. The initial experimental conditions had been only poorly reported, with no peer review before publication. As other researchers rushed to duplicate the results that Pons and

The Day the World Changed

Fleischmann claimed to have observed, group after group soon announced negative results. Though today a few die-hard researchers continue their investigations into cold fusion, the overwhelming scientific consensus is that cold fusion is not valid, and that the initial, amazing results from Utah arose from mistakes in setting up the experiment and in recording the data from it. The scientific community has concluded that the entire episode highlights the danger of rushing to make a premature announcement of an unproven phenomenon.

Seven years later, on August 7, 1996, no one ignored the possibility that the news of possible ancient life on Mars might repeat the cold-fusion episode, leaving in its wake embarrassed scientists and a confused public. The team of researchers not only recalled the story of cold fusion but also well knew that three decades earlier, scientists had reported finding other structures hinting at life, including pollen grains, inside a meteorite found near Orgeuil, France, implying that life might well exist in interplanetary space. As this startling news spread around the world, Edward Anders of the University of Chicago, along with other skeptical scientists, investigated the Orgeuil meteorite further—and showed that the structures indicative of life were contaminants, forms of life on Earth that had somehow gotten inside the meteorite.

Fully cognizant of the Orgeuil meteorite story, the scientists studying the rock from Mars had carefully considered the possibility that it had been contaminated during either its years of repose in the Antarctic ice or the period when it had been collected, stored, and analyzed. Although they could not completely rule out this sort of contamination, they had taken every effort to avoid defiling their samples, and had performed comparison tests to show that blank samples of ordinary rocks, subjected to the same procedures, did not reveal what the Martian meteorite did. The use of such "controls" forms an essential part of any scientific experiment. In this case, the control experiment allowed the meteorite to pass the "cold-fusion test," by showing that whatever the researchers had found in the rock from Mars indeed came from Mars. What remained was the question of interpretation, the determination of the *meaning* of the evidence in the Martian meteorite.

The press conference, therefore, initiated an intricate, multifaceted debate, loaded with implications about the origin of life on Earth and throughout the universe, which has gone on for months since the discovery and will continue for years to come. The scientists participating in this debate all agree that the rock from Mars carries evidence about long-vanished eras on our planetary neighbor. They emphatically *dis*agree over the implications of that evidence—a fact that will provide much interest throughout the pages of this book.

No matter what the outcome of this ongoing debate, the announcement made on that day in August has already changed the way that both the scientific community and the public approach the fundamental questions raised by the search for extraterrestrial life: Where did we come from? How universal is the story of life on Earth? Are we alone? Or does the cosmos teem with life, both like and unlike our own? And what can we do to answer these questions?

The fact that we may already have samples of possible ancient life on Mars in our own laboratories has changed deeply ingrained paths of thinking about these issues, whether or not we eventually find that life did exist on Mars, or does exist there now. "This is the first evidence we've found that a planet like Earth had life-inducing chemistry," says Tobias Owen, an expert on planetary science at the University of Hawaii. "The fact that this evidence exists in the *first* rock discovered from the era when Mars had conditions favorable to life implies that prebiological activity, if not life itself, was common long ago on Mars—and thus is likely to have been common on Mars-like and Earth-like planets throughout the universe. Before August 1996, those of us who studied the possibilities of extraterrestrial life were in a domain close to religion: We believed that it must exist, but had no real evidence. Now we've got evidence."

Furthermore, the meteorite from Mars, more than three times older than any Mars rock we had previously found—and thus the single example of a Martian rock from truly ancient times—has a significance quite independent of its implications for extraterrestrial life. "The rock shows that Mars carries a crucial record that is completely *missing* from Earth," says Owen. "That is the record of the era when life began here on Earth, somewhere between 3.8 billion years ago, the ages of the oldest rocks we have, and 4.5 billion years ago, the time that the sun and its planets began to form. This fact deepens the incentive to return to Mars, in order to begin to examine that ancient record on the planet itself."

Why does Mars retain a geological record of its earliest years, while the Earth does not? Plate-tectonic motion, the slow grinding of pieces of the Earth's crust against and over one another, has buried all the rocks and debris from erosion that are more than 3.8 billion years old. "Those are the years when life appeared on Earth," says Christopher McKay, an expert on Mars who works at the NASA/Ames Research Center in California. McKay (no relation to the leader of the team that investigated the Mars meteorite) calls the span of time between about 3.8 billion and 4.5 billion years, which includes the time when life began on Earth, the "Wonder-bread years" after a food once advertised as important to growing children. "Mars has a record of its 'Wonder-bread years,'" McKay says, "because it has so much less erosion and plate-tectonic activity than Earth." Cursed or blessed with a size much smaller than Earth's, Mars lacks the internal heat sources that on Earth nearly melt the layer of rocks just below the crust. As they slide on this semifluid layer, pieces of the crust wander slowly over the underlying mantle of our planet, occasionally causing earthquakes as they brush against or dive under one another. These plate-tectonic motions sooner or later bury the individual sections of the crust, depriving us of any fossil record of life that it may contain.

We know that during long-vanished eras, three to four billion years ago or more, the environment on Mars was far more favorable to life than it is now. In those bygone epochs, Mars apparently had streams and lakes, with flowing and still water, rainfall and evaporation, whose record still remains, stamped on the apparently lifeless planet. Today Mars has no liquid water whatsoever; any ice that warms up on its surface immediately becomes water vapor. "The *Vikings* arrived a few billion years late," Christopher McKay notes, referring to the

This mosaic of photographs of Mars, taken by the *Viking* spacecraft in 1976, shows the canyon system called Valles Marineris, which has depths up to five miles and extends for more than 2,000 miles across the planet's surface. Mars's diameter, 4,220 miles, equals just 53 percent of Earth's. (*NASA photograph*)

two NASA spacecraft that reached Mars in 1976. "And they literally just scratched the surface in their search for life."

Each of the two *Viking* landers picked up small samples of Martian soil and studied them for evidence of some form of life. Despite early excitement over what seemed to be positive signals, the *Viking* biology teams concluded that the surface of Mars today appears to be devoid of life. Christopher McKay feels confident that a better exploration of Mars will reveal clear signs of ancient life, and possibly life even now that lies hidden below the surface. "The best candidates [in a reconnaissance for Martian life] are ancient lake beds," he says. "Think of the Bonneville Salt Flats in Utah, which are also an old lake bed. You drive a rover around the ancient lake, and then right up the bed of the creek that once fed it. That's where you drill to look for life." We shall see later how and why the conditions favorable to life on Mars evolved into an environment hostile to life, and the experiments that could test for microscopic forms of alien life. ("We'll know it when we see it," says Chris McKay.) This issue of identifying extraterrestrial life lies at the heart of any discussion of the evidence found in the meteorite from Mars.

What Did the Scientists Find in the Rock from Mars?

The results announced at the press conference on August 7 and published in *Science* magazine on August 16, 1996, can be grouped under four main headings:

The Day the World Changed

- The rock contains globules of *carbonates*, mineral deposits made from carbon and oxygen atoms combined with other atoms such as calcium, iron, or magnesium. About one percent of the rock's mass consists of these globules, each of them not much more than a hairsbreadth wide. On Earth, creatures living in the oceans produce carbonate deposits (a good thing for us, since this process keeps carbon-dioxide gas from dominating our atmosphere and warming our planet beyond what we can tolerate), but carbonates can also form in the absence of life. A central part of the debate over the Martian meteorite asks: When and how did these carbonate globules form? What does their formation imply about conditions on Mars? In particular, did these carbonates form in the presence of liquid water, like most carbonates on Earth?

- Throughout the carbonate globules in the Martian rock, sophisticated techniques reveal molecules called PAHs (polycyclic aromatic hydrocarbons), which living organisms often produce as they decay. Although PAHs can arise from nonbiological processes as well as from organic matter, the PAHs within the globules are distributed unevenly, implying that tiny organisms might have produced their concentrations.

- Within the carbonate globules, the scientists also found crystals of two types of magnetic minerals, iron oxide and iron sulfide, that are in essence minuscule magnets. On Earth, some bacteria manufacture similar minerals, which they use to sense "up" and "down" and thus to orient themselves as they float, seeking regions more favorable to their metabolism. The production of iron oxides and iron sulfides requires an environment less acidic than that needed to form carbonates. Hence their presence *within* the carbonates suggests that the magnetic minerals might have been made by living organisms that maintained, as bacteria often do on Earth, two separate situations in close proximity.

- Finally, powerful electron microscopes revealed that the rims of the carbonate globules contain tiny structures, elongated "ovoids," whose shapes resemble those of the smallest single-celled bacteria found on Earth. These tiny bacteria, among the smallest forms of life on our planet, have sizes measured in microns. Each micron is one one-thousandth of a millimeter, just 0.00004 inch. Fifty to one hundred "microbacteria" a micron wide would be needed to span the diameter of a human hair. As small as these terrestrial microbacteria are, they are five to a hundred times larger than the ovoids found in the meteorite from Mars.

None of the scientists who participated in this research and appeared at the press conference thought for a moment that the four items of evidence *prove* the existence, beyond all doubt, of ancient life in the Martian meteorite they had studied. They well knew—and would be repeatedly and vociferously reminded during the months to come—that carbonate deposits can be made without the presence of life; that PAHs have been found in other meteorites, and probably in interstellar space as well, without anyone claiming that life exists there; that magnetite and iron sulfide might well have been made in a different environment from the carbonates, then deposited within them; and that the ovoids might in fact be crystalline structures made without the intervention of life. "None of [our] observations is in itself conclusive for the existence of past life," the researchers had written in their paper for *Science* magazine.

Finally, David McKay, Everett Gibson, and their colleagues at NASA and elsewhere ended their paper with a sentence so completely unexceptionable as to verge on the disingenuous. "Although there are alternative explanations for each of these phenomena taken individually, when they are considered collectively, particularly in view of their spatial association, we conclude that they are evidence for primitive life on early Mars." Of course they are *evidence*, other scientists responded immediately, but *how good is that evidence?* The stage was set for an ongoing drama: the trial of the Martian rock.

The Courtroom of Science and the Burden of Proof

If we use the analogy of a criminal trial, we can see at once that the scientists must not only serve as expert witnesses but also play the roles of advocates, arguing for and against the evidence for ancient life on Mars. The prosecution lawyers seek life on Mars, while the defense team argues that life has not been proven. Along with the scientists who follow the proceedings, we ordinary citizens must play the role of the jury. Who, then, plays the judge, whose function, in American courtrooms, consists mainly of ruling that particular items of evidence are admissible or inadmissible? In real life, this job remains unfilled, so I shall take it on, not as completely as a judge would, but still always ready to offer comments on the scientific evidence. With the roles assigned, we lack only the rules. Among these we must choose one of decisive importance: What standard does the courtroom of science impose on the evidence in order to reach a verdict?

In a case such as this one, held in an unfamiliar jurisdiction, any attorney worth a shingle would immediately demand to know how thoroughly the prosecution must demonstrate that no explanation other than life can be sustained. Must this occur beyond a reasonable doubt, the standard required in criminal cases? Or must the prosecution prove its case simply by a "preponderance of the evidence," the standard imposed in disputes between two individuals? Perhaps we should use a third, intermediate standard such as "compelling evidence," whatever that may be. Furthermore, if we impose a reasonable-doubt standard, we raise additional, related issues. Must the prosecution disprove *every* alternative explanation, or only those that the defense introduces? And must every alternative, no matter how unlikely, itself be disproven beyond a reasonable doubt?

Since we have no science courts, these questions have no general answers, even though they arise every day among scientists, who must convict or acquit theories that attempt to explain the phenomena that other scientists observe. In deciding whether or not ancient life existed on Mars, you the jury must attempt to understand the attorneys' presentation of evidence and their arguments for and against a particular conclusion. This may appear daunting. However, to those who have followed the evidence presented and argued in great criminal trials, the task of learning and judging the evidence for and against life on Mars should prove comparatively simple, and certainly offers a much greater emotional payoff.

Suppose that we imagine two great teams of attorneys who will argue for and against a verdict of life on Mars. We can begin the trial with the stipulations, the items that any two reasonable attorneys (after all, the attorneys are actually scientists) can and will agree are true. These stipulations aim at creating a list of *what must be proven in order to establish the existence of life.* For example, one key item must be the demonstration that all the exhibits entered into evidence have undergone the proper procedures concerning their discovery, the subsequent chain of control over them, their later analysis, and (most difficult of all to establish) their correct interpretation. We can therefore specify the following steps as requirements for proving that life once existed on Mars:

(1) Establishing that a rock on Earth in fact came from the planet Mars.

(2) Demonstrating that the material inside that rock did not receive contamination from terrestrial forms of life during the thousands of years while it awaited discovery, while it was being collected, or at any time thereafter.

(3) Proving, by detailed analysis of matter in the rock, that it contains chemical compounds and structural forms that are characteristic of life.

(4) Eliminating the possibility that this chemical and

structural evidence could have arisen through processes *other* than those performed by living creatures.

Point (4) conforms to the well-known dictum laid down by Sherlock Holmes, the great, though entirely fictional, detective created by Arthur Conan Doyle. Holmes, who lectured his companion Dr. Watson unmercifully, restated his number-one rule of criminal investigation to Watson in *The Sign of Four*: "How often have I said to you that when you have eliminated the impossible, whatever remains, *however improbable*, must be the truth?"

This principle proves useful in locating the car keys that you have dropped somewhere around the house, but it suffers in scientific practice from the fact that not everyone will agree on what constitutes an alternative explanation, or on whether that explanation has been truly eliminated. Nevertheless, every scientist depends on it to establish the validity of results obtained at the price of much hard work. "Scientists are great at *disproving* things," says Richard Zare, the final author (such is life among the Zs) of the *Science* article. "What we are *not* good at is *proving* things." Hence the need to follow Holmes's dictum: Identify everything that needs disproof, proceed to disprove it, and your case stands as proven.

In the case of life on Mars, the four-fold task that we have imposed on the prosecution grows progressively more difficult. Suppose that Lawyer A argues for the defense (no life on Mars) and Lawyer Z for the prosecution (life on Mars does exist). We shall find that A will readily admit point (1), that the rock came from Mars; will probably concede point (2), that contamination did not occur; will grapple heartily with point (3), allowing that the rock does contain some *evidence* for life on Mars; but will dig in his heels and utterly refuse to concede point (4), that alternatives other than life should or must be rejected. Attorney A may argue that a greater burden of proof should be imposed on some of the items on the list than on others. For example, if we cannot reject the possibility of contamination beyond a reasonable doubt, it hardly seems worthwhile to proceed through the rest of the trial. Attorney Z will likewise gladly concede that point (4) contains the crux of the case, and that in order to prevail, he must show that all possibilities other than life can be rejected. But must these possibilities be rejected beyond a reasonable doubt, as Attorney A may insist, or only by a preponderance of the evidence? If Lawyer Z can show by a preponderance of the evidence that an explanation based on life is more likely than all the alternatives, that would be quite a startling result—even if the evidence cannot surmount a beyond-reasonable-doubt standard.

Enough of this thorny issue of the burden of proof, which each of us must assess and resolve. Let us review the stipulations that the attorneys might reasonably agree upon:

(1) A meteorite found in Antarctica in 1984, named ALH 84001 by its cataloguers, was identified in 1993 as having originated on the planet Mars. The conclusion follows that some impact must have blasted this rock loose from Mars, and that it eventually collided with the Earth after orbiting for years through interplanetary space.

(2) The meteorite had been carefully collected with a view to avoiding any contamination of its surface, and its interior apparently underwent no contamination from the terrestrial environment during its approximately 13,000 years in the ice. The total age of the rock, including its time on Mars and the much shorter interval that it spent in interplanetary space, amounts to 4.5 billion years.

(3) Detailed examination of the interior of this meteorite revealed carbonate globules. Although the date when these compounds formed cannot be established with precision, some evidence suggests an age of 3.5 billion years, while another analysis points to 1.4 billion years. Within these carbonate globules, scientists have found other minerals and compounds that living creatures generate. Most controversially, they have also found tiny ovoids, structures that *may* be fossils of long-vanished creatures. Debate continues over whether anything this small would have been *too* small ever to have been alive. The burden of proof rests on those who assert the ovoids are fossils to demonstrate the merits of this assertion.

(4) The carbonate globules in the meteorite contain molecules of the type called PAHs, as well as magnetic minerals of two different types, iron oxides and iron sulfides. Although all of these compounds can be explained without invoking the presence of living creatures, the fact that they all appear in the same regions cannot find quite such an easy explanation. The debate over whether the alternative explanations can be eliminated therefore centers on how likely it

is that nonbiological processes could have generated all of the compounds found within the globules.

Human Fascination with Mars

If NASA and President Clinton had announced that a rock from an asteroid contained tantalizing hints of ancient life, too small to be seen without employing the most powerful electron microscopes, most of us would have told them to call back when they had something better. But Mars is different. Along with Halley's Comet, the rings of Saturn, and the Man in the Moon, the "Red Planet" has earned a special place in the public consciousness, capable of inspiring both fear and awe in culture after culture.

For as long as human beings have stared at the night sky, they have crafted myths about the bowl of night, the sky that seems to rotate around the Earth, carrying with it myriad points of light. Among the host of fixed stars, which maintained the same positions with respect to one another as the sky appeared to turn, observers soon noted five bright objects meandering among the starry constellations. These *planets*—which means "wanderers" in Greek—constantly changed their positions with respect to the stars, slowly moving through the constellations of the zodiac. Although these motions occur too slowly to be seen on any one night, a keen-eyed observer could see that a planet such as Mars and Venus moves slightly away from its star-studded position of the previous night. We now know that these changes arise from the planets' orbital motions around the sun. Of the five planets visible to the naked eye, Venus comes closest to Earth and shines with the greatest brightness, but since Venus always remains relatively close to the sun as we see it in the sky, we never observe the planet in the full depth of night.

Blood-red Mars, sometimes almost as bright as Venus, follows no such rule. This planet passes through a full cycle in its appearances above the horizon. Mars sometimes rises and sets with the sun, but about a year afterward, Mars will rise at sunset, spend all night crossing the sky, and set only with the dawn. As Mars approaches its greatest brightness and stays up nearly all night, its wandering against the starry background seems to cease for a few nights. Mars then *reverses* its direction of wandering with respect to the stars, as Earth, which moves more rapidly in its orbit, overtakes the red planet, so that at its maximum brightness, Mars appears to move backward with respect to the starry background, like a telephone pole seen from a fast-moving car against the backdrop of a distant landscape. After a few months of this retrograde motion, Mars changes its direction of wandering once again and resumes its usual direction of motion through the constellations. Jupiter and Saturn exhibit similar behavior, but in a less pronounced fashion. In addition, these two planets change their brightnesses far less than Mars does, because the distance from Earth to Mars undergoes far greater changes than the distance to Jupiter or Saturn.

Captivated by Mars's red color, by its strange motions, and by its changing brightness, cultures throughout the ancient Middle East associated this planet with the unpredictable, temper-prone god of war. Mars represented the Babylonian war god Nergal and the Greek war god Ares. The Romans, adopting Greek thought even more slavishly than the Greeks had appropriated Babylonian ways, made a modest name change from Ares to Mars, which we still use. More than a millennium after the fall of the Roman empire, Johannes Kepler plotted the orbit of Mars in his successful attempt to discover his laws of planetary motion, and found that Mars's orbit around the sun has an elliptical shape, not a circular one as previous authorities had insisted. In *Gulliver's Travels*, written during the early eighteenth century, Jonathan Swift imagined two moons for Mars; a century later, astronomers discovered that Mars indeed possesses two small moons, Phobos and Deimos, which are captured asteroids only a few miles across.

At the end of the nineteenth century, Percival Lowell, a high-caste Boston brahmin, built an observatory near Flagstaff, Arizona, for the sole purpose of studying Mars. Motivated by Giovanni Schiaparelli's reports from Italy of "canali"—lines or channels—on Mars, Lowell proceeded to map dozens upon dozens of Martian "canals." Starting in 1895, Lowell trumpeted his conclusions in popular, well-written, and well-received books. Mars has highly intelligent inhabitants, said Lowell, who have dealt with an arid planet by channeling melting water from the polar caps to the equatorial regions. For half a century, Lowell's observations seemed reasonable, at least to those who had not fully investigated the tendency of the human eye and brain to visualize lines between dark patches on a light background. In 1898, one of the first great science-fiction novels, H. G. Wells's *War*

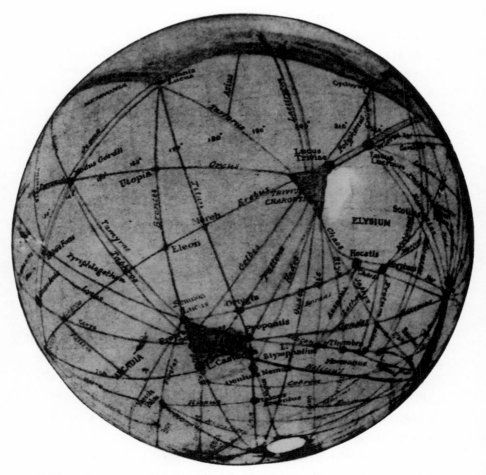

Nearly a century ago, Percival Lowell drew this map of Mars and what he thought were its canals. In Lowell's view, the Martian canal network extended all around the planet's surface, converging at what he took to be oases. (*Courtesy of Lowell Observatory*)

The Day the World Changed

of the Worlds, built on Lowell's work to imagine a technologically advanced race of Martians who invade Earth to harvest its resources and eventually succumb to earthborn germs. Forty years later, on Halloween eve, October 30, 1938, Orson Welles's Mercury Theatre broadcast a radio dramatization of Wells's novel, provoking a widespread panic along the East Coast that brought millions of Americans face-to-face with the realization, celebrated afterward in hundreds of science-fiction movies, that a horde of alien invaders would leave humanity with no place to hide.

Three decades after the night of false news from Mars, during the late 1960s and early 1970s, NASA launched four *Mariner* spacecraft to Mars that could send photographs back to Earth. The *Mariner* images laid to rest the notion that Mars showed signs of an intelligent civilization. Confronted with arid reality, Lowell's canals vanished; gone too was the notion of liquid water on Mars; gone was the possibility that Mars's surface might have easily detectable life. In its place, especially after the *Viking* missions of 1976, there appeared a cold, dry Mars, hostile to life, perhaps not only without life today but also with not the least apparent history of life.

In one day, the Martian meteorite ALH 84001 largely altered this attitude. Whether or not it demonstrates that Mars has or had life, the 4.5-billion-year-old rock from the next planet out eliminated a crucial absence in a long-standing scientific search: For the first time, we now have a sample from a planet other than our own that shows conditions favorable to life. Tobias Owen compares the news from the Martian rock with the first picture transmitted from space to show the whole Earth. "Its message will sink in slowly," says Owen, "but it will change the way that we think about life in the universe." If we want to let the news wash over us, we must begin by understanding the history of the rock that made headlines around the world.

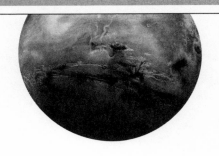

"I'd always

known

that rock

was weird."

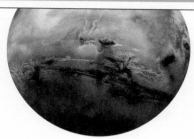

THE STORY OF
A ROCK

All the evidence announced in 1996 as pointing toward ancient life on Mars lay within one four-pound, three-ounce rock, discovered in the Allan Hills of Antarctica 4.5 billion years after the rock had formed and 13,000 years after it had fallen to Earth. To find this rock from Mars, along with thousands of other meteorites, cost less than one million dollars, expended on the National Science Foundation's Antarctic meteorite search program. This amount, a fraction of a percent of the cost of sending a spacecraft to Mars to bring samples back to Earth, serves to remind us that although we can and should pursue some of the marvelous, well-planned, complex projects that offer us great scientific returns, we should also remember to notice what we may stumble across. "The most difficult thing to discover is something you're not looking for" is the way that the cosmologist Rocky Kolb expresses this phenomenon.

Since a team of expert searchers discovered the Martian meteorite as part of a careful, continuing search for meteorites in Antarctica, where meteorites on the ice can be identified with relative ease, we can hardly say that it was found by accident. What *was* accidental was the fact that this particular meteorite turned out to have come from Mars and also to be by far the *oldest* of the dozen Mars rocks found on Earth. The rock carries the scientific name ALH 84001, meaning that it was the first meteorite listed among those found in the Allan Hills of Antarctica during 1984. Scientists' diligent efforts, coupled with the joys of serendipitous discovery and the ability to profit from it, have now established certain basic facts about the oldest rock we possess from the planet Mars. Under the stipulation procedure that we envisaged for our imaginary teams of scientific, rational lawyers, we can tell the story of the rock without objection from either side.

ALH 84001: Four and a Half Billion Years of History

ALH 84001 formed on Mars about four and a half billion years ago within a mass of volcanic material that slowly cooled and solidified. Most of this volcanic rock is ortho-pyroxene, a silicate rock made primarily of silicon and oxygen, mixed with some iron and magnesium. About

The Story of a Rock

ALH 84001, by far the oldest rock from Mars found on Earth, measured about six inches long by three inches high and weighed in at just over four pounds. The small cube to the lower right of the rock has sides that are one centimeter—.3937 inch—long. (*NASA photograph*)

half a billion years after the rock had hardened, some four billion years ago, an object colliding with Mars near the end of an era of intense bombardment struck this rock and melted part of its material once again. Four billion years later, just 15 or so million years ago, another impact blasted the rock loose from Mars. After spending nearly all of those 15 million years orbiting the sun, the rock collided with Earth about 13,000 years ago: a mere yesterday in the history of the solar system, since 13,000 years amounts to less than $1/1000$ of the time that the rock spent in space, and less than $1/300,000$ of the time since cooling lava first produced the rock on Mars.

The Hunt for Life on Mars

How can scientists determine these time intervals and trace the different stages of ALH 84001's journey through space and time? Even though some of the data listed above may not prove to be completely accurate, the evidence behind each item seems convincing to those who have spent years studying the strange careers of meteorites. Both the amount of time that the meteorite spent orbiting the sun and the interval since it struck the Earth can be determined from the fact that objects moving through interplanetary space are continuously struck by particles called *cosmic rays*.

Cosmic-ray particles, which are produced when stars explode and blast their outer layers into space, are mainly protons, electrons, and helium nuclei, all traveling at nearly the speed of light. Cosmic rays permeate space and bombard any object not shielded from their effects, either by the metal walls of a spaceship or by a blanket of atmosphere such as our own. Even Mars's thin atmosphere offers protection against cosmic rays, but once in space, any object receives a cosmic-ray blitz. This bombardment produces new types of atomic nuclei in the object at a steady, well-measured rate—until the object falls into the protection offered by Earth's atmosphere. We now know the rate at which cosmic rays generate new atomic nuclei, and we also know the rate at which these nuclei decay into other types. As a result, by measuring the numbers of nuclei made by cosmic rays and also the numbers of their decay products, meteorite specialists can determine how long the object traveled through interplanetary space and how long ago that journey ended. Measuring the effects of cosmic-ray bombardment and subsequent atmospheric protection resembles determining the amount of time a man has spent out in the rain and then indoors by finding out how thoroughly soaked his clothes became and how much time has passed since they began to dry. This approach revealed that ALH 84001 passed about 15 million years in the rain of cosmic rays, followed by 13,000 years of essentially zero cosmic-ray bombardment.

What of the largest time intervals associated with the rock from Mars, its age of 4.5 billion years and the evidence of a strong impact on it 4 billion years ago? These ages come from a technique called *radiometric dating*, which involves measuring the numbers of atomic nuclei that have been produced by radioactive decay of some of the elements inside a rock such as ALH 84001. Before we examine this technique further, we must examine the larger issues of where meteorites come from and what they have to tell us, once we can recognize and analyze them.

Stones from Heaven

Our planet undergoes a continuous bombardment by rocks from space. Although this fact has been well established, we can easily understand why men of good common sense long ago rejected this notion, despite tales from faraway lands of stones that had fallen from heaven. To this day, the holiest spot of Islam, the cubical building called the Ka'aba, houses the Black Stone of Mecca, whose veneration predates the origin of Islam in the seventh century A.D. The Black Stone's darkness suggests a meteoritic origin, though legend makes it a gift to Adam upon his expulsion from Eden, and explains its color as the result of a gradual change from white to black through the stone's absorption of the sins of countless pilgrims.

Old ways of thinking have a long persistence: Three centuries ago, when European scientists began to suspect that some of the strange rocks found on Earth's surface had indeed come from above, they had to overcome the lingering medieval belief that the cosmos is made of a fifth substance, a "quintessence," entirely different from the four substances—fire, air, earth, and water—that form our world. If that were so, we could hardly expect heavenly material to consist of rocks, or of mixtures of rock and metal, that appear similar to rocks on Earth. Besides, how could stones fall from heaven without our noticing them?

Edmond Halley was the man who first suggested an extraterrestrial origin for the objects we call meteorites. Halley, best known for his studies of a comet that passes relatively close to Earth every 76 years, analyzed observations of meteors, familiarly known as "shooting stars," that different observers had recorded at various locations in England. From these observations, he calculated that bright meteors, seen simultaneously from sites separated by dozens of miles, must have velocities of many miles per second. To Halley, these enormous speeds implied that the objects could have been formed "by some fortuitous concourse of atoms" in regions beyond the Earth. Despite their respect for Halley, few of his fellow scientists took his hypothesis seriously until more direct evidence became available long after Halley's death in 1742.

On July 24, 1790, several hundred people witnessed a spectacular shower of meteors near the town of Agen in southwestern France. Immediately afterward, strange-looking rocks were collected nearby, and were described in a scientific journal whose editor judged the reports of stones falling from the sky to be just plain impossible. In the same year, the Abbé Andreas Stütz, an Austrian naturalist, stated that "in our time it would be unpardonable to regard such fairy tales as likely." However, just four years later, Ernst Chladni, a well-known physicist who had founded the science of acoustics, published his analysis of hundreds of reports on meteors and meteorites and strongly favored the hypothesis of extraterrestrial origin, particularly for the iron-rich meteorites. Though Chladni's work gained favorable attention, it convinced only a minority of those who debated the origin of meteorites.

More than a decade later, on December 14, 1807, an extremely bright meteor passed over New England, accompanied by a fall of objects near Weston, Connecticut, not far from Yale University. Yale's professor of chemistry, Benjamin Silliman, and the college librarian, James Kingsley, gathered chunks of the new-fallen material, one of which weighed as much as a man. An old and apocryphal story states that President Thomas Jefferson, one of the greatest friends of science ever to occupy the White House, attempted to apply the "Sherlock Holmes rule" by stating "It is easier to believe that two Yankee professors would lie than that stones would fall from heaven."

The Hunt for Life on Mars

By the time of the Weston fall, careful observers many miles apart who recorded meteors' trajectories at different angles above the horizon had used triangulation to show that most meteors appear a few dozen miles above the Earth. When these observations were supplemented with actual findings of strange rocks after a particularly bright meteor had passed, the conclusion grew ever more reasonable that some of the objects seen at altitudes of many miles had passed all the way through the atmosphere to reach the Earth's surface. A clinching event occurred near L'Aigle, France, on April 26, 1803, when a bright meteor appeared in clear skies, producing violent detonations and leaving behind thousands of newly fallen meteorites. As reliable accounts of this event spread throughout the scientific community, meteorites seemed likelier to be stones from heaven.

On November 13, 1833, what is now called the Leonid meteor shower reached astounding proportions: Observers in the eastern United States reported seeing many meteors per minute. This stimulated further research, and astronomers soon demonstrated that meteor showers recur on an annual basis, though some years may bring far more spectacular showers than others.

Once scientists recognized that meteors come from space, the reason for the annual recurrence of meteor showers became evident. Meteors arise because hosts of small objects, which scientists now call meteoroids, orbit the sun on their own trajectories, creating meteoroid "swarms." If a meteoroid happens to encounter the Earth, its velocity with respect to our planet typically equals many miles per second. The meteoroid's high-speed collision with the upper atmosphere heats it enormously: The gases cannot get out of the meteoroid's way and pile up in front of it, compressed by the ram force of the object. The same effect occurs when astronauts return to Earth in a spacecraft; in both cases, the ram pressure heats the object as it pushes through the atmosphere. Spacecraft must be carefully designed to minimize the effects of this heating as they enter the atmosphere at speeds of "only" a few miles per second. Meteoroids, traveling still faster and without control systems, heat to the point that they glow. We then see a moving point of light, which we call a shooting star or meteor, as the meteoroid evaporates, either partially or completely. The shooting stars visible on any clear night far from city lights usually are objects no larger than a pebble, seen several dozen miles overhead, so hot that they produce a brief glow, the last flash of their existence.

Some swarms contain especially large numbers of meteoroids. In that case, once every year, at the time when the Earth crosses a swarm's orbit, we can expect to see a large number of meteors. Several times each year we have the chance to watch a few good nights of a particular meteor shower, such as the Quadrantids, which reach their peak about January 3; the Perseids, with a maximum on August 12; and the Leonids, which peak around November 16. The names of these swarms denote the constellations from which the meteors appear to radiate, a detail that depends on the direction with which the Earth is moving with respect to the swarm of meteoroids. If the Earth happens to intersect a swarm at a point that is particularly rich in meteoroids, the resulting meteor shower will be especially intense, as happened with the Leonids in 1833.

The Hunt for Life on Mars

An unusually bright meteor, capable of causing you to inhale sharply, ask "What was that?" and dig your fingers into your companion's arm, arises from an object perhaps half an inch across, evaporating 30 to 50 miles above you. If the object has the size of a grapefruit, it can create a "fireball," an intensely bright meteor that leaves a trail behind it for a few seconds; some of an object this size may survive its passage through the atmosphere, becoming a meteorite as it hits the surface. To bring 200 pounds of material to Earth, as in the meteorite that hit Connecticut in 1807, requires an initial object with a mass considerably larger than the surviving part found on Earth. Every day hundreds of tons of meteoritic debris strike the Earth, almost all of it in the form of particles weighing much less than an ounce. A tiny fraction of these meteorites weigh a quarter-pound or more, and are sufficiently large to be noticed by a careful observer.

As the nineteenth century turned into the twentieth, sharp-eyed observers continued to find meteorites. Some meteorites consist mainly of rocks, not noticeably different from rocks on Earth, while others are mixtures of rock and metal, and still others mostly metal. Since metal-rich material on the Earth's surface catches the eye much more readily than strangely pitted rocks do, the metal-rich meteorites proved far easier to recognize than stony meteorites. In 1897, Admiral Robert Peary brought a large metallic meteorite from Greenland to New York, where it remains on display in the Hayden Planetarium. At 36 tons, this object holds the record for a meteorite that has been moved; its only superior in bulk, several times more massive, lies partially buried in Namibia at the site where it landed. Small meteorites are not scarce. Popular-astronomy magazines carry advertisements from hardworking collectors, offering them for sale at reasonable prices. But what of the larger ones?

Meteoritic Impact with Earth

When a multiton meteorite strikes the ground at speeds of many miles per second, the local effect must be devastating, though the worldwide impact remains insignificant. Even the meteoroid that produced the famous Meteor Crater in Arizona, almost a mile across, when it struck the Earth about 50,000 years ago probably did nothing to affect our Neanderthal ancestors living in what we now call other states and foreign countries. This object probably was "only" the size of a bungalow and weighed a mere 20,000 tons; most of it fragmented upon impact or lies buried beneath the crater it produced. Throughout past aeons on Earth, similar objects have probably struck our planet at intervals measured in tens of thousands of years.

Now imagine the effect on Earth of an impact from a meteoroid the size of San Francisco. This "meteoroid" would in fact be either an asteroid—basically just a large meteoroid—or a comet, a dirty snowball made mainly of ice and frozen carbon dioxide, not quite as dense as an asteroid of equal size but still capable of inflicting enormous damage on Earth. When it struck our planet, a ten-mile-wide object would shake the Earth. If it happened to strike in the seas, the asteroid or comet would thrust miles of ocean water out of its path at hypersonic speed, so that its effect would be the same as if it hit one of the continents. In either case, the impacting object would open a ten-mile-wide hole in the atmosphere—and in the oceans too, if the strike occurred there. During the minutes that the atmosphere and seas would take to refill the area emptied by the onrushing object, an enormous amount of fine grit, dust, and even larger particles would rise through the hole, blasted from the Earth's surface by the force of the impact. Carried to altitudes of dozens or hundreds of miles, the larger particles would envelop the Earth in seething swarms of hot, glowing pebbles, which could set the world's forests on fire as they fell back to the surface. The smaller particles, most dangerous of all, would spread all around the world, coating the atmosphere with black soot, returning only slowly to Earth's surface, floating for months or even years because of the particles' small sizes, like an oil slick on ocean water. The soot would create a planet-wide darkening that would last until the air gradually managed to cleanse itself. The darkening in turn would wreak havoc with the lives of all living creatures, save those few that inhabit the ocean depths or live in mile-deep rock crevices.

This scenario once seemed only a fantasy, designed to demonstrate what might happen if space were not so large that celestial objects almost never collide. During the past two decades, however, the impact theory has now become well established as a description of the most famous of all extinctions on Earth. Sixty-five million years ago, an asteroid ten miles or so in diameter struck what we now call the Yucatán Peninsula in Mexico. The fossil record demonstrates that this epoch coincided with a *mass extinction*, a time when a significant fraction of all the Earth's species disappeared within a few hundred thousand years, or possibly far more rapidly. The organisms that vanished during the mass extinction 65 million years ago included all the species of dinosaurs, which had stood atop the food chain for more than a hundred million years while our shrewlike mammalian ancestors hunted for insects and grubs. (Since many biologists now regard birds as a surviving branch of the Dinosauria, we may note that the mass extinction killed only all the *conventional* dinosaurs.)

Paleobiologists have not yet established a causal link between the giant impact and the demise of the dinosaurs 65 million years ago. However, the coincidence in time, together with worldwide evidence that the object that struck the Earth then sent some of its material into the stratosphere, from where it settled back to Earth, has convinced most of those who study mass extinctions that this one arose from a great collision with our planet. Other mass extinctions, which have occurred at roughly 30-million-year intervals during the past few hundred million years, may have arisen from the same sort of impacts—or from quite different causes. In any case, the dinosaur extinction speaks eloquently of what stones from heaven can do to Earth, every once in a rare while. Fortunately, larger meteoroids are rarer than smaller ones, and objects large enough to be called asteroids are rarer still. If we want to obtain a large sample of stones from heaven, we must search for meteorites of modest size.

The Happy Hunting Grounds for Meteorites

Meteorites rank among the rarest of all the categories of rocks on Earth. Precisely because their chemical composition differs markedly from that of most rocks at the Earth's surface, meteorites are more subject to weathering and degradation from exposure to the conditions on our planet. These two facts make searching for meteorites a difficult, time-consuming affair, one that no one engages in as a full-time occupation. A single place on Earth violates the common-sense rule against letting your children grow up to be meteorite hunters: the Antarctic.

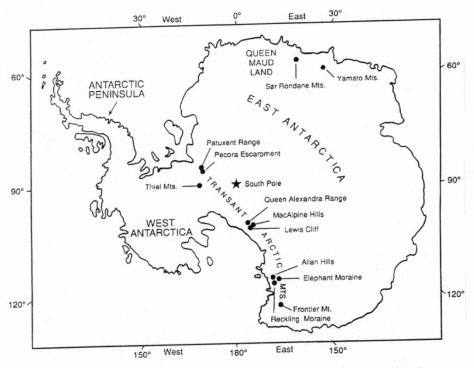

This map shows the location of the major "meteorite fields" in Antarctica. McMurdo Station, the chief scientific outpost on the continent, lies on the seacoast, close to the four Allan Hills meteorite fields, the Elephant Moraine, and the Reckling Moraine. (*Courtesy of Meteoritical Society*)

The Story of a Rock

The south polar land mass, twice as large as the forty-eight United States, consists mainly of a two-mile-thick ice sheet that covers the continent below. In precipitation terms, most of the Antarctic is a desert. Except for the band near the ocean that surrounds Antarctica, the continent receives less snow each year than Washington, D.C. What falls, however, tends to remain. Temperatures that may reach a few degrees above freezing on the warmest summer days in January fall as low as minus 140 Fahrenheit in winter, providing a natural deep freeze that inhibits the weathering of any meteorites.

Even better for meteorite finders, the region of Antarctica where the Trans-Antarctic Mountains protrude through the ice contains isolated peaks and ridges called "nunataks" that funnel the ice sheet as it slowly slides downward toward the sea. Between the nunataks and the seaward ends of the ice sheet, at six thousand feet or so above sea level, lie regions of "blue ice," where strong winds continually sweep the ice free of snow. Some of these blue-ice plateaus are the happy hunting grounds for meteorites, because the slow movement of the ice sheet around the nunataks tends to concentrate the meteorites spatially. In addition to the dozen or so fields close to the Trans-Antarctic Mountains, two other blue-ice fields exist on the other side of the continent, in Queen Maud Land, where large topographical barriers also divert the direction of the ice's slow flow.

During the past three decades, the blue-ice fields have yielded more than ten thousand meteorites. Each of these specimens now bears a one- to four-letter identifying symbol, followed by a number that gives the year of discovery and the order in which the meteorite was listed in the laboratory. MAC 86002 is the second meteorite found in 1986 at the Macalpine Hills, and Y 88141 is the 141st meteorite found in 1988 in the Yamato Mountains of Queen Maud Land, where more than six thousand meteorites have been discovered. Nearly two thousand meteorites bearing the designation ALH have been found in a rich site close to a spur of the Trans-Antarctic Mountains called the Allan Hills. This site consists of four sub-areas—the Main, Near Western, Middle Western, and Far Western Icefields. The first searches in the Allan Hills, from 1976 through 1978, concentrated on the Allan Hills Main Icefield, and later years these efforts were extended to the other three fields. In 1981, the Allan Hills Middle Western Icefield yielded ALH 81005, the first meteorite identified as coming from the moon. Three seasons later, a still greater treasure turned up in another icefield near the Allan Hills.

How Robbie Score Found the Ancient Rock from Mars

ALH 84001, the meteorite that rocked the world, was found on December 27, 1984, in the Allan Hills Far Western Icefield, about a hundred miles from McMurdo Station, the chief scientific outpost on the polar continent. On the day of this discovery, Roberta Score was part of a seven-person team searching for meteorites. Like mountaineers, such a group properly regards its finds as the product of a team effort, but fame has its rules, and the score went to the single woman on the team.

The Story of a Rock

Besides, Roberta "Robbie" Score had earned her chance at fame through years of familiarizing herself with Antarctic meteorites: She had been working with these meteorites at the Johnson Space Center since 1978. Born and raised in Detroit, Score had had enough of Michigan after her freshman year at Michigan State; she left the university and moved to Los Angeles. "I hung out as a dental assistant," she remembers, "and then went back to school at UCLA to become a dentist." To satisfy a science requirement, Score enrolled in a geology course, and before long had changed her career goal to geology. "When I graduated, I got a job at the Antarctic Meteorite Center in Houston—and stayed sixteen years." In the fall of 1996, Score moved to a new job: supervisor of the

In the fall of 1984 Roberta Score was on her first trip to "the ice" when she found the chunk of rock named ALH 84001. (*Photograph © Barry Staver*)

chief research facility in Antarctica, the Crary Laboratory, named in honor of the scientist Albert Crary, the first person to visit both the North and South Poles. The Crary Lab is the pride of McMurdo Station, whose population swells beyond one thousand during the Antarctic summer season (November through February). "I am recognized as the hometown celebrity," she reports. "Those I work with love to embarrass me by announcing to anyone who will listen that I was 'the one.'"

Score's timing in beginning her job as meteorite hunter proved to be excellent. Japanese scientists had found the first meteorite in Antarctica in 1969, but the first wide-ranging Antarctic searches began only in 1976 with a joint United States–Japanese effort. The program lasted for three years, with the researchers dividing their finds (literally in the case of the larger meteorites, one by one for the smaller ones) for study in their home laboratories. Since the Johnson Space Center had developed elaborate facilities to analyze rocks from the moon, it was the natural center for the United States' efforts in studying Antarctic meteorites, which, after 1978, separated completely from the Japanese efforts. "I'd been at the Johnson Space Center since the beginning," Score notes; "the [meteorite] laboratory has my personality."

In the fall of 1984, which was spring in Antarctica, Roberta Score was on her first trip to "the ice," as Antarctic hands call the continent for reasons that are strikingly apparent to even a casual visitor. Early in December, helicopters airlifted the team of meteorite hunters, along with all their food and equipment (including a snowmobile for each of the seven members) to the Allan Hills, where for seven weeks they camped on the ice every night—not that night ever falls in December or January at that latitude. "By December 27, we'd found about a hundred meteorites," Score recalls. "We were searching the Far Western Icefield," a region about 30 miles long and 5 miles wide, and some 45 miles west of the Allan Hills Main Icefield. "When we started out that day, it was cold and windy, but it got calmer and warmer. We were in our normal search mode—on our snowmobiles about 30 meters apart—when we saw 'ice pinnacles,' which are thought to be produced by slow-moving ice and ferocious winds. At last, some scenery in a flat landscape! Some of the pinnacles were 15 feet high. They're associated with crevasses, so there was also a sense of danger. And then as we were getting back to work, I ran into this rock. I did what we always do: got off the snowmobile and waved my arms to attract attention. Finally, everyone came over. We put a flag in the ice to mark the location and got out the special clean bags from the Space Center. You do your damnedest not to contaminate the rocks. This rock had a real green appearance; I picked it up by wrapping a bag around it, flipped the rock inside the bag, taped the bag shut, wrote a field number on the tape, and put it in the box."

Of the three hundred or so meteorites collected on that trip, Score remembers the green one best. The collection, packed into ice chests, made the journey northward from McMurdo Station by the end-of-season ship, whose major cargo is that year's accumulation of trash and debris, brought back to the U.S. for recycling. From Port Hueneme, California, where the ship docked, the rocks were flown in dry ice to Houston. There the meteorite experts placed them within a nitrogen chamber to dry them out, photographed and characterized each one as to its apparent type, gave them laboratory numbers, and removed a small chip from each meteorite, which they sent to the Harvard–Smithsonian Center for Astrophysics in Cambridge, Massachusetts. Score was amazed to find that her "green" meteorite was in fact a dull gray, its "color" apparently the result of the Antarctic lighting and the dark glasses she had been wearing. "I'd kept saying, 'Wait till you see this rock!' And then everyone looked at it and said, 'Yeah, yeah.' " Nevertheless, since it was her job to assign laboratory numbers, Score gave this rock the pride of first place: ALH 84001, the first meteorite recorded from the 1984–85 search effort. Just over four pounds in weight, the size of a small potato, about three and a half times as dense as water, with an outer dark crust that showed the modification caused by its high-speed passage through the atmosphere, Score's find was destined to become the most famous of all the rocks that have fallen to Earth.

How the Rock Gained Its Fame After Remaining Obscure for a Decade

Safely stored in its nitrogen-filled cabinet at the Johnson Space Center, ALH 84001 was initially classified, on the basis of a single chip analyzed under a laboratory microscope, as a diogenite, a relatively common type of meteorite, thought to have been knocked into interplanetary space from the large asteroid Vesta by collisions with smaller objects. It was not until 1993, nine years after its discovery, that ALH 84001 came under the scrutiny of David "Duck" Mittlefehldt, a geologist working as a staff scientist for the Lockheed Martin Engineering and Sciences Company in Houston, which had a contract to provide technical and scientific support to the Johnson Space Center.

"I was interested in comparing different types of meteorites," Mittlefehldt recalls, "so I took a close look at a thin section of ALH 84001. What tipped me off was that this rock had significant amounts of trivalent iron oxides [iron-oxygen compounds in which the iron atoms each share three electrons with neighboring atoms]. Diogenites tend to have all their oxidized iron in divalent form [sharing only two electrons with their neighbors]. I got another thin section, and looked at the sulfides [the compounds that sulfur forms with other atoms]. The iron sulfides were disulfides [two sulfur atoms for each iron atom] instead of monosulfides [one sulfur per iron atom]. Then it all clicked."

As a meteorite expert, Mittlefehldt had in effect recognized a Mars rock by its gestalt. At that time, eleven meteorites had been identified as Martian; ALH 84001 would be the twelfth. What was missing was a fingerprint to clinch the identification. "After I found the iron disulfides, I found carbonates [compounds of oxygen and carbon, together with calcium, iron, or magnesium]. Then I told Marilyn Lindstrom, the meteorite curator, that I was sure the rock was Martian. She was excited, but she said we had to get an isotope analysis to be sure. So we sent a sample to Robert Clayton at the University of Chicago to get a count on the oxygen isotopes. He got results identical to those for other Martian meteorites."

ALH 84001 thus made an extremely short list of rocks known to have come from Mars. "I was thrilled!" says Roberta Score. "I'd always known that rock was weird."

Dating Rocks from Mars: The Fingerprint of the Isotopes

Of the tens of thousands of meteorites collected on the Earth's surface, experts have now shown that ten came from the moon and a dozen from Mars. To be able to make this statement required years of patient effort to find and identify a cosmic fingerprint that marks the origin of this tiny minority of all meteorites, and also furnishes us with their ages.

This fingerprint resides in the details of the different types of atomic nuclei in the meteorites. Every atom consists of electrons that orbit a central nucleus made of protons and neutrons. The number of protons identifies the *type of atom* (six for carbon, seven for nitrogen, eight for oxygen, and so on), while the number of neutrons specifies the particular *isotope* of that atom. For example, all carbon atoms have six protons in each nucleus, but the carbon isotopes differ in the number of neutrons that each atom contains, which may be six, seven, or eight. Scientists specify each isotope by adding the number of neutrons to the number of protons, so carbon's three isotopes are carbon-12, carbon-13, and carbon-14.

All the isotopes of a particular type of atom display the same chemical behavior: They interact in identical fashion with other atoms, because they all have the same number of protons and electrons, which determines their chemistry. Only at the nuclear level of interactions, when one nucleus collides with another, does the number of neutrons, and hence the type of isotope, make a difference. Nuclear interactions occur only in situations of extreme violence, typically involving enormous temperatures, in which high-energy collisions strip each nucleus bare of its surrounding cloud of electrons. All the different types of nuclei—all the isotopes of each particular type of atom—originate in violent collisions, either in the early universe, in stars, in stellar explosions, or, as we have seen, in interplanetary and interstellar space, when high-energy particles called cosmic rays collide with dust and rocks.

Some isotopes are stable, capable of enduring without change for billions of years. Not surprisingly, these isotopes dominate the universe. Other isotopes are unstable, and undergo radioactive decay: They change into other types of nuclei, on time scales that vary from a fraction of a second to billions of years, depending on the type of unstable isotope. Carbon, for instance, has two stable isotopes, carbon-12 and carbon-13. The third isotope, carbon-14, is unstable: These nuclei are "radioactive," and decay to produce nuclei of nitrogen-14.

Although the radioactive decay of any individual nucleus cannot be accurately predicted, the results of these decays among large numbers of nuclei can be described both statistically and accurately. We know, for example, that carbon-14 nuclei decay with a *half-life* of 5,750 years. The half-life denotes the interval needed for half of the nuclei in any large set to decay. After two half-lives have elapsed, only one-quarter of the original number of unstable nuclei will be present; after three half-lives, only one-eighth, and so on. Eventually, nearly all of the unstable "parent nuclei" will decay to produce "daughter nuclei," and the ratios of parent and daughter isotopes provide a set of whorls on the cosmic fingerprint of isotope ratios. If we can determine the original numbers of parent and daughter nuclei in an object, and can measure their present numbers, we can calculate how many half-lives have elapsed since the object formed. The decay of carbon-14 nuclei with a half-life of 5,750 years allows scientists to perform "carbon dating," which they use to assign ages to old timber and fabric. For example, this technique has dated the famous Shroud of Turin to the thirteenth century A.D.—a blow to those who imagine this to have been Jesus' burial shroud.

Every element (all atoms with a particular number of protons and electrons) has its own set of isotopes, some stable and others unstable. Carbon-14 dating works well for ages measured in thousands of years, but for much longer intervals, so many half-lives have passed that essentially no carbon-14 remains to be detected. To apply the radioactive-decay technique over longer time scales, we must look to unstable isotopes with much longer half-lives. One of the best isotopes for long-term dating is potassium-40, an unstable nucleus that decays into argon-40 with a half-life of 1.28 billion years. Suppose that a rock forms from volcanic material, whose heat drives off all the argon, which is an inert gas that can escape by bubbling through the molten rock. In that case, the epoch when the rock cools sufficiently to be a rock will mark the last time that the rock has parent isotopes (potassium-40) but no daughter isotopes (argon-40). Afterward, by measuring the relative amounts of parent and daughter, we can determine the age since the rock solidified. Equal amounts of parent and daughter would imply an age of 1.28 billion years, whereas three times as much of the daughter as the parent would imply twice that age, or 2.56 billion years.

In the case of ALH 84001, the method based on the comparison of the potassium-40 and argon-40 abundances yielded an age of about 4.0 billion years. This apparently represents not the time when the rock first formed, but rather a time when an impact melted at least part of it. We can draw this conclusion because another dating method provided the basic age of the rock, 4.5 billion years. This age was found for a sample of ALH 84001 by the geologist Emil Jagoutz and his colleagues at the Max Planck-Institute for Chemistry in Mainz, Germany, who compared the amounts of rubidium-87, the parent isotope, and strontium-87, the daughter isotope into which the parent decays with a half-life of 47 billion years. The most reasonable interpretation of these two ages seems to be that the rock first solidified 4.5 billion years ago, but was melted in part half a billion years later, releasing all the argon-40 and resetting the clock provided by the decay of potassium-40.

The eleven meteorites previously identified as originating on Mars all had ages between 170 million and 1.3 billion years. Thus *ALH 84001 is more than three times older than the next oldest rock from Mars.* Since plate-tectonic motions have buried all Earth rocks older than 3.8 billion years, and since the oldest lunar rocks have ages of 4.2 billion years, ALH 84001 stands out as by far the oldest rock we have ever obtained from a planet or a planetary satellite. However, meteorites, which rank among the earliest objects that collected themselves together as the solar system formed, have ages comparable to that of the ancient Martian rock. This might suggest that ALH 84001 has somehow been misclassified, and in fact belongs to the class of ancient but otherwise ordinary meteorites that never saw the surface of Mars, but the identification of ALH 84001 with an origin on Mars appears to be certain.

How to Recognize Rocks from Mars

Like the age-dating methods, the technique for locating the place of origin of certain strange rocks found on Earth depends on measurements of the amounts of different isotopes. In this case, however, the cosmic fingerprints that characterize a rock and allow us to identify where it came from reside in the numbers of *stable* isotopes, not the isotopes that undergo radioactive decay. Almost all elements have at least two stable isotopes, such as carbon-12 and carbon-13. Nitrogen's stable isotopes are nitrogen-14 and nitrogen-15; oxygen's are oxygen-16, oxygen-17, and oxygen-18; and neon's are neon-20, neon-21, and neon-22. The ratios of the numbers of isotopes of a particular element vary, though not by enormous amounts, from place to place in the cosmos. Astronomers who measure isotopic abundances in stars find that most stars show ratios similar to the sun's, though unusual stars often have one or more elements whose isotopes appear in uncommon ratios.

The Story of a Rock

In our solar system, two of the sun's four inner planets, Earth and Mars, have atmospheres made of gases whose isotopic ratios have been measured. In both cases, a single set of ratios characterizes the entire planet, because the atmosphere spreads all around the planet as a well-mixed blanket of gas. The Martian atmosphere consists mainly of carbon dioxide, with small amounts of nitrogen, water vapor, carbon monoxide, and other gases such as argon, krypton, and xenon. The two *Viking* spacecraft that landed on Mars in 1996 sent back a detailed inventory of the amounts of these elements and the relative abundances of their different isotopes. We know, for example, that the ratio of nitrogen-15 to nitrogen-14 has a larger value on Mars than on Earth, and have measured isotope ratios for a dozen other elements on Mars.

The meteorites identified as Martian show isotopic ratios unlike those on Earth but just like those on Mars. One of these meteorites, EETA 79001, found in the Texas Bowl of the Elephant Moraine a few dozen miles from the Allan Hills, actually contains glassy inclusions that enclose small portions of Mars's atmosphere—as verified by the isotopic ratios within this gas. David McKay has called this meteorite a "kind of Martian Rosetta stone," because it provided definitive proof of its own Martian origin. The meteorites that closely resemble the Rosetta stone meteorite may then be judged to be Martian as well. This resemblance to EETA 79001 lies in the meteorites' isotopic ratios: They have all incorporated sufficient material from the Martian atmosphere to provide the isotopic fingerprint that says, This glove fits the owner; the rock once belonged to Mars.

The Hunt for Life on Mars

In 1976, the analysis of the composition of the Martian atmosphere made by instruments on the *Viking* landers was the responsibility of Tobias Owen, now at the University of Hawaii. "One of our finest measurements was the determination that an isotope of xenon, produced by the decay of radioactive iodine, is about two and a half times more abundant on Mars than on Earth," Owen says. "To find the same pattern of isotope abundances on Mars in the tiny amounts of xenon trapped in these meteorites certainly argues for their Martian origin. The relative abundances of other inert gases in the meteorites support this conclusion, since they are different from the pattern in our atmosphere, on Venus, or in any other meteorites, but match those found on Mars."

Even the water in the Martian meteorites is unearthly: Each gram of water contains as much as five times more deuterium, the rare, heavy isotope of hydrogen, than the water that we drink on Earth does. This high deuterium abundance exactly matches the value that Owen and his collaborators found in the water vapor in Mars's atmosphere by spectroscopic analysis from the Mauna Kea Observatory in Hawaii. Still more proof for the Martian origin of the meteorites came from an examination of the oxygen atoms bound into the silicate minerals within them. When Robert Clayton, a meteoriticist at the University of Chicago, measured the ratio of oxygen-18 to oxygen-16 nuclei in ALH 84001 and found a value different from the ratio in Earth's atmosphere but identical to the Martian number, all doubt about the rock's place of origin vanished.

The ten meteorites from the moon have been identified by using the same technique of comparing isotopic ratios, which in this case means juxtaposing the ratios in the meteorites with those in rocks brought from the moon to Earth. Small but significant differences between the lunar and terrestrial isotopic ratios have convinced experts that at least some of the moon never belonged to Earth. The current favored theory for the origin of the moon envisages an immense collision, four and a half billion years ago, between a Mars-sized object and the Earth-in-formation. This collision spewed material from Earth into nearby space, where it mixed with matter from the colliding object, as well as with matter that had not yet joined any larger object, to form our satellite.

Meteoriticists (a tongue-twisting name for those who analyze meteorites) refer to rocks from Mars as "SNC meteorites." SNC stands for Shergotty, Nakhla, and Chassigny, three places where these rocks have been found. Six of the twelve SNC meteorites were found in Antarctica; the remaining six, including S, N, and C, fell to Earth on other continents. Five of the twelve weigh in at one pound or less; five others weigh between four and eighteen pounds, and two, Zagami and Nakhla, are considerably larger, with 40 and 90 pounds of matter. The most massive of the SNCs, the Nakhla meteorite, fell on the town of this name in Egypt on June 28, 1911, killing a dog. (In 1954, a woman in Alabama was hit by an eight-pound meteorite that came through the roof and struck her in the leg; but this was a garden-variety rock, not a Mars meteorite.) All of the SNC meteorites consist of types of rock familiar to geologists, typically basalts formed from volcanic outflows. Their unique ratios of isotopes stamp them as rocks from Mars.

Getting a Piece of the Rock: The Antarctic Meteorite Committee

Who owns the rocks from Mars? And who allocates samples from them to researchers around the world who seek to probe these meteorites? Answering the first of these questions could provide long-term employment for teams of lawyers, all of whom would recall the very first case they studied in law school courses on real property, the case of the disputed fox. In 1805, the Supreme Court of New York adjudicated the lawsuit brought by a sportsman named Post, who had "started" a fox for his dogs to chase, against Pierson, another sportsman who had caught it, and ruled that the rights of property concerning wild animals are "acquired by occupancy [what we would now call possession] only," so that Post had no legal right to the fox. Building on this firm foundation of precedent, and analogizing wild animals to untamed meteorites on the ice, the attorneys could argue that ALH 84001 belongs to Roberta Score, or to the team of seven meteorite hunters.

As a practical matter, all the meteorites found by the United States Antarctic Meteorite Program are under the control of a ten-member committee called the Meteorite Working Group, whose members represent NASA, the National Science Foundation (NSF), the Smithsonian Institution, and various universities. The Smithsonian Institution has been designated as the national curator of all meteorites that land on federal property or are discovered with federal funding, and the National Science Foundation has been instructed by Congress to fund and control all of the United States' scientific research within

30 degrees of the South Pole. Currently headed by Ursula Marvin, a geologist at the Smithsonian Astrophysical Observatory (part of the Harvard Center for Astrophysics in Cambridge, Massachusetts), the meteorite committee meets twice each year to allocate samples of the Antarctic meteorites. Technically the committee only advises the Smithsonian and the NSF, but its recommendations have always been followed.

The meteorites themselves are stored at the Johnson Space Center, except for small pieces that have been chipped from them, which reside at and belong to (at least in some sense) the Smithsonian. The seven-ounce sample from ALH 84001 that NASA displayed at the August 7 press conference was the Smithsonian's piece of the rock. Another seven ounces had already been distributed to researchers, including the portion that David McKay and his collaborators had analyzed. Ursula Marvin reports that when the committee met on the last weekend of September 1996, and considered the 53 ounces that remained from ALH 84001's original 67, "we decided to hold a moratorium before allocating any more samples." In placing a hold on any further distribution, the meteorite committee anticipated that both NASA and the NSF will call for proposals from scientists, will then ask a committee to review them, and will provide some governmental funding to the successful applicants, to whom the committee will presumably agree to yield pieces of the ancient rock from Mars.

"We never allocate everything," notes Marvin, meaning that the committee always makes sure that parts of every meteorite remain available for further study. Marvin anticipates that the NSF and NASA will

also encourage further studies of other Martian meteorites, of the ancient carbonaceous-chondrite meteorites,
and of the oldest rocks on Earth. Comparisons within
and between these groups of rocks will help to place the
news from ALH 84001 in proper context. And just what
is that news? In hopes of resolving this issue, we must
return to the courtroom of science, and ask both teams
of attorneys to make their best case for and against ancient life on Mars.

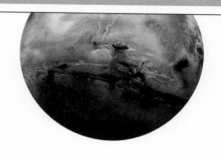

"Would you

like to

look at an

interesting

meteorite?"

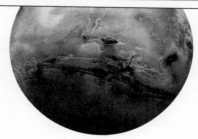

THE EVIDENCE
IN THE ROCK

We have seen that meteorite experts agree that we can have some confidence in our knowledge of the history of ALH 84001, which includes the rock's volcanic origin on Mars four and a half billion years ago, its almost equally long residence on or just under the Martian surface, its sudden expulsion into space 15 million years ago, and its collision with the Earth some thirteen thousand years before Roberta Score found it in 1984. When we look inside this rock, we enter a region in which experts debate mightily—not over all the relevant evidence, but most strongly over the items that point toward possible ancient life. To draw full value from these experts, we must prepare to hear their testimony and to enjoy their debates, which

ALH84001,0

To study the Martian meteorite ALH 84001, scientists cleaved the rock with diamond saws to produce samples for detailed analysis. The small cube to the lower right has sides one centimeter long. (*NASA photograph*).

reveal the strong and weak points of their conclusions and the evidence on which the interpretations rest.

The Carbonates of ALH 84001

No one was surprised to find that most of ALH 84001 consists of silicates, which are compounds of silicon and oxygen and form the dominant component of the rocks on Earth. The types of silicates, called orthopyroxenes, that form the bulk of the oldest planetary rock ever discovered closely resemble the composition of many silicates, young and old, on Earth. In 1993, what David Mittlefehldt *did* find surprising in his examination of a sample from ALH 84001 was the fact that this volcanic rock is rather rich in carbon compounds, and most notably in carbonates.

Carbonates are mineral deposits made from carbon and oxygen atoms, bound together with calcium atoms (as is true for most carbonates on Earth) or with atoms of magnesium, iron, manganese, or other elements. When Mittlefehldt made a careful analysis of a sample from ALH 84001 in his laboratory at the Johnson Space Center, he saw small globules of carbonate scattered throughout the silicate rock. The largest of these globules measured about a hundredth of an inch in diameter, and the smaller ones only a few thousandths of an inch. Like all scientists, Mittlefehldt measures small distances not in inches but in microns (millionths of a meter) and nanometers (billionths of a meter). An inch covers just over 25,000 microns, each of which contains a thousand nanometers. If you care to split hairs, each of the finer strands spans about 50 microns, so a nanometer equals 1/50,000 of a fine hair's diameter. The analysis of the meteorite from Mars and all the commotion that arose from it deal with the details that appear at the micron and nanometer levels of size.

On Earth, carbonates almost always form in the presence of water, and are often associated with beds of fossils in limestone (carbonate-rich) rocks that formed from the slow accumulation of the shells of countless tiny sea creatures. Although nonbiological processes can produce carbonates, the carbonates made from living organisms are even more common, such as the limestone rocks that constitute the White Cliffs near Dover, England. Examining the carbonate globules in ALH 84001, Mittlefehldt saw some with spherical shapes, others flattened nearly into disks. What impressed him most, however, was the fact that nearly one percent of the total mass of ALH 84001 consists of carbonates, an unusually large amount. What were these carbonates doing in an ancient rock formed from volcanic material?

Mittlefehldt sought the advice of Chris Romanek, an expert in the chemical processes that form and change carbonates, who was also a Lockheed Martin employee working under contract to the Johnson Space Center. "Duck came across the hall to my laboratory one night and asked, 'Would you like to look at an interesting meteorite?' and showed me the carbonates," Romanek remembers. "Then I went to a science meeting where researchers working on terrestrial carbonates explained how they had used acid to etch away some of the material to allow closer study of the remainder. So I proposed that we do some etching, and we saw internal structure in the carbonates. We took the pictures we'd made—they seemed very unusual to me—to Everett Gibson, a geologist at the JSC. Gibson said we had to look at the carbonates in greater detail. Down the hall, David McKay was in charge of a microscopy lab. Gibson showed him the pictures, and he and McKay formed a research team to work harder on the meteorite."

In 1994, Romanek, Mittlefehldt, Gibson, and other geologists wrote a paper for the British science journal

Nature, the approximate equivalent of *Science* magazine in the United States. "We concluded that the carbonate globules had formed at temperatures between 0 and 80 Centigrade," Romanek says. Because 0 and 100 Centigrade mark the freezing and boiling points of water, the formation of carbonates between 0 and 80 C meant that liquid water could have existed. And since most of the carbonates on Earth have formed underwater, with or without the presence of living organisms, the temperature that Romanek and his colleagues had determined strongly suggested the presence of liquid water on Mars at the time of formation of the carbonate globules in ALH 84001. The fact that the oldest rock from Mars contained evidence for liquid water at the time when the carbonates had formed engaged no one's interest more than David McKay's.

David McKay and the Carbonate Globules

David McKay, who became the leader of the team that studied ALH 84001 and published the paper in *Science* magazine, is an expert geologist and geochemist, born in 1936 in Titusville, Pennsylvania, a locale geologically renowned for providing the oil-bearing strata that allowed the first oil well in America to begin production a century and a half ago. When McKay was ten, his father, who worked for the Kewanee Oil Company, moved his family from Pennsylvania to the more promising oil regions of Oklahoma. Already interested in becoming a geologist by the time he entered high school, McKay went to Rice University as an undergraduate, received a master's degree in geochemistry at the University of California at Berkeley, and then worked for Exxon for a year, an experience that motivated him to return to Rice to obtain his Ph.D. in geology with an emphasis on geochemistry. In 1964, back in Berkeley as a National Science Foundation postdoctoral fellow, McKay received an offer to work at NASA's Manned Spaceflight Center in Houston, a name eventually changed to the Johnson Space Center. McKay returned to Houston as NASA was gearing up for the human exploration of the moon and has now completed three decades of employment there.

Between July 1969 and December 1972, six *Apollo* missions (numbers 11 through 17, except for *Apollo 13*, whose misfortunes eventually made a gripping film) brought lunar samples back to Earth, as did three un-

David McKay, the geologist who led the team of scientists who investigated ALH 84001, has worked at NASA's Johnson Space Center for three decades. (*Photograph by Donald Goldsmith*)

manned spacecraft sent by the Soviet Union. Most of the samples from the *Apollo* program ended up in Houston, where McKay, who had helped train the astronauts in how to recognize interesting pieces of stone, had access to the first rocks from another world. "Those were exciting times," McKay remembers. "There was lots of speculation about the nature of the [lunar] soil." This included considering whether living organisms might exist in the rocks and dust from the moon, despite the fact that the moon receives intense ultraviolet radiation from the sun, lacks both an atmosphere and any form of water, and exhibits nothing that resembles the sorts of changes that life might produce on an object's surface. Even though the probability of lunar life seemed extremely remote, and even though any such life probably could not survive on Earth, scientists quite understandably worried about even an infinitesimal chance that life from the moon could harm our planet. Fortunately, the moon rocks have been examined around the world without loosing any lunar plagues on an unsuspecting humanity.

Throughout the 1970s, as the *Apollo* missions ended but the analysis of the lunar samples continued, McKay's detailed studies made him one of the world's experts on the characteristics of the lunar surface, especially concerning the particles called agglutinates. "These are unusual particles, forming the majority of some parts of the lunar soil, that were made by impact from micrometeorites," McKay says, adding, "I've studied agglutinates until I'm sick of them."

When McKay learned that David Mittlefehldt had demonstrated the Martian origin of ALH 84001, "I put together a small consortium in our division to study it further. When I studied the thin sections [slices of rock made with diamond saws, so thin that light shines through them and reveals their contents], I was intrigued by the carbonate globules, which looked unusual. I wanted to understand how the globules formed. Now, the meteorite tends to crumble slightly when you cut it. Presumably it's crumbling along the network of shock fractures we see in it. It's not really a soft rock, though; the minerals are quite tough. We think most of the carbonates formed within some of those fractures." McKay and his colleagues set up their equipment to probe inside the carbonate globules, hoping to determine their structure and composition. They were, of course, aware that an ancient rock from Mars might have incorporated fossils or other indications of life. "Everett Gibson and I were both interested in signs of past biological activity," says McKay, "so we agreed to team up."

The Ovoids at the Edges of the Carbonate Globules

During the summer and fall of 1994, McKay, Gibson, and the other researchers at JSC sectioned their sample of ALH 84001 for examination, first with conventional microscopes and then with much more powerful electron microscopes. To probe a hair-wide carbonate globule requires the most advanced electron microscopes, which "see" fine details by means of electron beams. Electron microscopes work by firing beams of electrons at a sample and noting what happens as the electrons reflect from or stream through the target. Scanning electron microscopes (SEMs) bounce electrons from a surface, while transmission electron microscopes (TEMs) shoot electrons through a thin sample. From the details of the electron bounces (in SEMs) and the diversions from straight lines of the penetrating electrons (in TEMs), electron microscopists can construct an image of the target. These images reveal the sample's structure with a precision of detail completely unobtainable with light waves, no matter how fine an optical system one may build.

"When we decided to look primarily at the carbonate structures," Chris Romanek says, "this brought in Kathie Thomas-Keprta, who's an expert on scanning and tunneling electron microscopy. With this expertise, we uncovered additional interesting structures." To examine small chips from ALH 84001, Thomas-Keprta and the other researchers at the Johnson Space Center used the highest-resolution SEMs in existence, capable of seeing details only a few nanometers in size—less than one ten-thousandth of a hairsbreadth. At this level of sizes, the meteorite from Mars becomes highly interesting. The JSC researchers found that most of the carbonate globules have orange colors. The rims of the globules, however, are nearly black and white in color, with the black rims on either side of the white regions, like an Oreo cookie. The carbonates in the black parts of the rims are rich in iron, while the white areas consist of magnesium-rich carbonates. At the globules' outermost rims, scanning electron microphotographs revealed a host of elongated structures, which the researchers called "ovoids," typically 20 to 200 nanometers long and 5 to 10 nanometers wide.

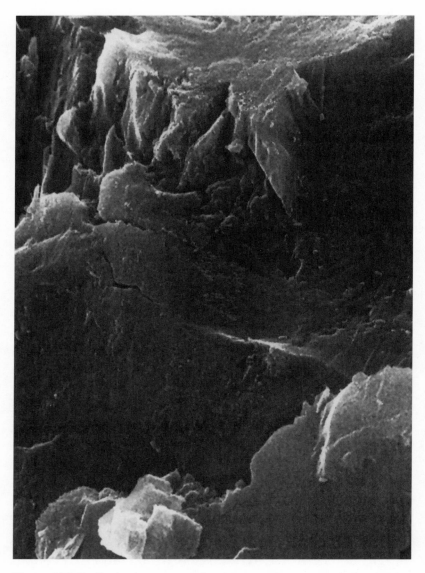

These "cliffs," which were cut by the saw that sectioned samples of ALH 84001, have heights measured in dozens of microns. Near the center of this image, just above a bright "shelf" of material, is an object a few hundred nanometers long, one of the objects whose ovoid shapes caught the attention of the geologists who studied thin sections of the Martian meteorite. (*NASA photograph*)

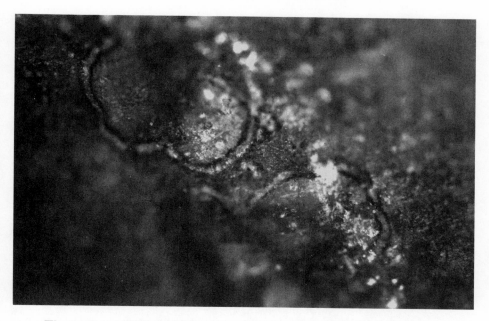

Throughout ALH 84001 are carbonate globules 100 to 200 microns in diameter such as those shown here. These globules have orange interiors, outside of which are whitish regions bordered by black rims. Some of the outer rims of the carbonate globules in ALH 84001 are rich in ovoid shapes, typically 50 to 100 nanometers in length and 10 or 20 nanometers wide. (*NASA photograph*)

To some of the JSC researchers, the ovoids' shapes resembled those of *microfossils*, the fossilized remains of bacteria so small that their sizes are measured in microns. But the ovoids at the edges of the carbonate globules are even smaller—smaller, in fact, than any known living organisms or microfossils on Earth. Does this mean that the ovoids are *too* small to have been alive? No one knows for sure. If living organisms had the size that the ovoids do, we could reasonably talk of *nanofossils*,

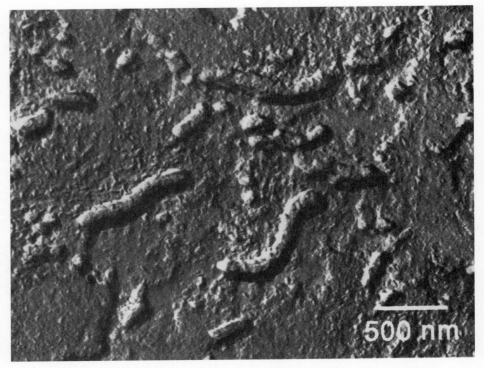

This transmission electron microphotograph shows some of the largest ovoid objects found in ALH 84001. The microphotograph was actually taken of a replica—a cast of a chip of the meteorite. To make this replica, the microscopists coated the area with a substance that solidified, then removed it and photographed it; this avoided decomposition of the features by the electron beam. The structure near the center of the image is nearly a full micron long. (*NASA photograph*)

fossils so small that their sizes are measured in hundreds of nanometers or less.

The fact that the ovoids have shapes that resemble those of terrestrial microfossils does not, of course, prove that the ovoids *are* fossils. McKay, Gibson, and Thomas-Keprta worked hard to show that these tiny objects with elliptical and ropelike shapes did not arise from contamination, but indeed belonged to the carbonate globules. In this they almost certainly succeeded, leaving the largest question of what left the ovoids behind when the carbonates formed. We shall address this question more fully in the next chapter; for now, we must turn to another key question: How old are the globules?

How Old Are the Carbonate Globules in ALH 84001?

ALH 84001 has an age of 4.5 billion years, but this only sets an upper limit on the age of the carbonate globules in the rock, which presumably grew through a process of accretion, the sticking together of minerals carried through fissures in the rock. If so, did the carbonate globules actually form as mineral deposits four billion years ago? Two? How can we tell?

On Earth, geologists set the ages of carbonate deposits by observing the fossil organisms within them and identifying the eras when those organisms lived. This method is not available for the rock from Mars (if it were, we would hear less discussion about whether life had existed there), so geologists turn to their favorite means of assigning ages—the decay of radioactive nuclei. Unfortunately, carbonates contain very little of those radioactive nuclei whose decay and decay products tell the story for other minerals. This goes double when one hopes to date a globule no larger than $1/100$ of an inch in diameter. The globules in ALH 84001 contain no measurable amounts of argon-40 or potassium-40, and very little of the rubidium-87 and strontium-87 that offer additional chances to apply this method. "It's a tough experimental problem, because the globules are so small," Chris Romanek notes ruefully.

Two different attempts at dating the globules have given two noticeably different ages. One preliminary result, cited by McKay and his colleagues in their article for *Science* magazine, gave an age of 3.6 billion years. As we shall discuss in Chapter 6, this would be just the right epoch to expect primitive forms of life on Mars. However, no one knows better than scientists how much we should beware leaping to conclusions because they fit our expectations and desires. "That's a very speculative date," says Richard Ash, one of the geologists who produced the 3.6-billion-year estimate. "It was almost a throwaway line in our published abstract," which summarized research still in progress, "and I'd be loath to say that this is a definite, good age for the carbonates [in ALH 84001]."

The other age assigned to the carbonate globules makes them far younger, a mere 1.39 billion years. Meenakshi Wadhwa, a geologist at the Field Museum in Chicago, and Guenter Lugmair, a chemist at the University of California at San Diego, derived this age from small chips of the meteorite, which they crushed and ran through magnetic systems that separated the iron-bearing minerals. Using a picking microscope, a magnetic separator, and acid to dissolve the carbonates away from the other material in the rock, Wadhwa isolated half a milligram of carbonate residue. About 50 parts per million of this amount consists of the strontium-87 nuclei that allow dating by the radioactive-decay technique that she and Lugmair applied to the sample.

The difficulties with this method of assigning an age lie not so much in the modest amounts of material with which the scientists must work as in the way they must interpret them. "Wadhwa's results depend strongly on the model of how the carbonates formed," Richard Ash notes. "You have to assume that all the carbonate material came from alteration of maskelynite," a type of glassy material formed by impacts. If the carbonate material came from other minerals than maskelynite, the 1.39-billion-year age becomes unreliable. "[Wadhwa's] actual measurements are very precise and her analyses are excellent," Ash says, "but they are subject to interpretation. There's always this problem: Is it the carbonate [you are dating] or the grains of maskelynite?" Wadhwa does not disagree. "The age is very model-dependent," she says, "and it remains to be verified. We hope to address this during our next few months of research."

The bottom line, Ash says, is that "getting ages for the carbonate globules is extremely difficult. Really, there are no truly reliable ages." Should Wadhwa's age for the carbonates prove correct, Mars would become still more surprising: If the carbonates formed in the presence of liquid water, as seems likely though not proven, Wadhwa's "young" age of 1.39 billion years implies that liquid water existed on Mars at that time. Yet current theories of Martian evolution, as well as a large amount of observational data, imply that Mars lost all its surface liquid water three billion years ago. Those of us with no particular axes to grind may conclude that dating the carbonate globules in ALH 84001 has not yet been successfully accomplished, and that the situation will probably not improve until geologists and geochemists obtain significantly more carbonate material.

Magnetic Minerals in the Carbonate Globules

Whatever the multibillion-year ages of the carbonate globules may be, it is the globules' contents that have caused the commotion. As part of the evidence for ancient Martian life that ALH 84001 contains, the researchers at the Johnson Space Center called particular attention to the fact that the carbonate globules contain two, possibly three, different magnetic minerals.

Most magnetic minerals are compounds of iron with oxygen or sulfur atoms. The minerals consist of individual "magnetic domains," each of which responds, like a tiny compass needle, to the Earth's magnetic field. What may come as startling news to the reader is the fact, known to specialists for two decades, that various forms of microscopic life on Earth, called *magnetotactic bacteria,* produce magnetic material for their own use. Many of these bacteria make a protein molecule called ferritin, which packs iron atoms in its core; a small fraction of them produce magnetic minerals, which are either magnetite, a compound of iron and oxygen, or greigite, made from iron and sulfur.

The magnetic minerals that were found in ALH 84001's carbonate globules are magnetite, pyrrhotite (whose composition resembles greigite's) and greigite itself, though this awaits further confirmation. All three of these magnetic minerals consist of molecules arranged in a repetitive, crystalline pattern. In fact, magnetite, an iron oxide, and greigite, an iron sulfide, have identical crystal structures: Magnetite consists of individual units made from three iron and four oxygen atoms, while greigite has units with three iron and four *sulfur* atoms locked into the same crystalline arrangement. Because iron atoms are responsible for the magnetic properties of all three minerals, they basically differ only in whether oxygen or sulfur atoms have combined with iron to form the mineral.

The magnetic mineral grains that Kathie Thomas-Keprta found in the sample of ALH 84001 have sizes of 40 to 50 nanometers; their roughly cubical and teardrop shapes resemble those of the magnetic minerals that terrestrial bacteria can manufacture, although they are not quite so uniform in size and shape as those in the bacteria. "When Dave [McKay] and Everett [Gibson] brought me into the project, I have to say I was the doubting Thomas," Thomas-Keprta told the August 7 press conference in a punning mood. "As time went on, I became more and more convinced [that the chemistry and structure of the magnetic minerals found in the meteorite's carbonate globules provide good evidence for ancient life on Mars]."

When bacteria make an iron oxide such as magnetite, or an iron sulfide such as pyrrhotite or greigite, they produce a host of tiny magnets, all with nearly the same size and shape. "These are masterpieces of permanent-magnet engineering," says Richard Frankel, a professor of physics at the California Polytechnic State University in San Luis Obispo. "You've got compass needles for one-micron organisms, about as perfect as you can make." The smallest bacteria that use magnetic minerals on Earth are only about one micron long—a mere 0.00004 inch, but still several times larger than the largest ovoids seen in ALH 84001. They use their magnetic minerals to sense the Earth's magnetic field, and then employ this information to discriminate between up and down. "The bacteria can't tell up from down by gravity," Frankel says, "but they live in a world with vertical chemical stratification, so knowing which way is up helps them get the nutrients they need." Like a scuba diver in murky water who cannot see anything distinctly but can still tell up from down, and can thus find his way to the surface, the magnetic bacteria have an advantage in orienting themselves for survival that their nonmagnetic cousins lack.

After Kathie Thomas-Keprta found magnetic particles in the carbonate globules of ALH 84001, David McKay and his collaborators at the JSC enlisted the aid of Hojatollah Vali, an Iranian-born biomineralogist (an expert on the minerals formed by living organisms) who teaches at McGill University in Montreal. Because of the low temperatures at which the carbonate globules had formed, Vali was surprised to find that they contained both an iron oxide, magnetite, and also iron sulfides, pyrrhotite and greigite. "You do find these together in nonbiological systems," he says. "Many hydrothermal situations produce both iron oxide and iron sulfide. But when you get them together, they usually have formed in a high-temperature situation." Thus, if the carbonates formed at relatively low temperatures, as Chris Romanek and his colleagues had concluded, the presence of both iron oxide and iron sulfide was difficult to explain on the basis of terrestrial experience. Furthermore, says Vali, "a solution that produces carbonates implies a low pH"—that is, a relatively high level of acidity. "But the formation of magnetite and iron sulfide implies a high pH [a non-acid, alkaline solution]."

Thus the discovery of grains of iron oxide and iron sulfide embedded in the carbonate globules implies two different environments, one needed to form the carbonates and one to produce the magnetic minerals. In addition, the existence of two types of magnetic minerals, one with oxygen and one with sulfur added to the iron, suggests a further splitting of environments. Almost by definition, living organisms manipulate environments to make a living, exploiting different environments to profit from the differences between them. Our stomachs, for example, maintain oxygen-poor conditions within which live billions of anaerobic bacteria that would quickly die if exposed to air. Hence finding magnetite and iron sulfides in the same place, and in a place where carbonates have formed, provides strong evidence that living creatures have passed that way. Since bacteria can manufacture magnetic minerals similar to those found in the carbonate globules, a biological origin for the iron oxide and iron sulfide in ALH 84001 seems reasonable, though hardly proven. "It's weird," says Joseph Kirschvink, a geologist at the California Institute of Technology, of the magnetic results in ALH 84001. "It doesn't mean [parts of the Mars rock] were alive, though. Maybe Mars is an extraordinary place."

Late in 1994, after many months of hard work on

the thin sections, with the electron-microscopy results and Vali's identification of the two magnetic minerals in hand, McKay and his collaborators began to wonder whether this meteorite might really have something. "We asked Bill Schopf to come out from UCLA to take a look," Chris Romanek remembers. "He's the expert on the signs of early life. Schopf made a visit and laid it on the line as to what would convince him. It wouldn't be enough to find structures that creatures might have made; we also had to find the organic matter associated with those structures." Since this was not the special province of the geologists at the Johnson Space Center, the researchers there had to look for experts at detecting organic molecules. "Kathie [Thomas-Keprta] had contacts at Stanford," says Romanek, "where they had the most sophisticated equipment and interpretation. So we started working with Richard Zare and Simon Clemett in the spring of 1995."

The Zarelab and the Discovery of Organic Compounds in ALH 84001

Richard Zare, a world leader in laser chemistry (the study of objects at the molecular level by means of laser beams), stands at the peak of his profession: He is not only a Stanford University professor but also the current head of the National Science Board, the oversight authority that supervises and sets policy for the National Science Foundation, the chief source of funding for scientific research in the United States. Born in Cleveland in 1939,

The Evidence in the Rock

Zare studied chemistry and physics at Harvard, entered graduate school at the University of California at Berkeley, and then returned to Harvard to follow his thesis advisor, Dudley Herschbach, who shared the chemistry Nobel prize in 1986 for his pioneering work in analyzing chemical reactions.

Armed with a Harvard Ph.D. at age twenty-four, Zare spent a year as a postdoctoral fellow at the University of Colorado and then became an assistant professor of chemistry at the Massachusetts Institute of Technology. Within nine months at MIT, his career hit a snag. ''I wanted access to the physics department's machine shop,'' Zare says. ''I was told that I could use only the

Richard Zare (right) and Simon Clemett (left) led the efforts at the Zarelab to analyze the composition of the carbonate globules in ALH 84001 without contaminating them. (*Photograph by Donald Goldsmith*)

chemistry department's shop, which was good for leather, brass, and wood—but I had to work with stainless steel. I went to the provost to complain, and he told me to have patience." Zare was not a patient man. "When I told the provost, 'I have an offer to go back to [the University of Colorado],' he told me, 'Young man, people do not leave MIT to go to Colorado.' The next day, I put my resignation on his desk. I got immediate permission to use the physics machine shop, but I never looked back. It was my first adventure—I switched from the chemistry department at MIT to the physics department in Colorado. Then the physics department sold half of me back to the chemistry department." A few years later, still shy of his thirtieth birthday, Zare was a full professor of chemistry at Columbia University.

"I was developing new techniques to look at small amounts of the products from chemical reactions while they are in the gaseous phase," Zare recalls of his time at Columbia. "That's where I first started laser ionization mass spectrometry"—the technique that would earn Zare fame in the world of physics and chemistry. In 1977, Zare allowed himself to be wooed to the West Coast by Stanford University, where he is now the Marguerite Blake Wilbur Professor of Chemistry and the head of an extensive laboratory, familiarly known as the Zarelab, that specializes in analyzing tiny amounts of compounds within a sample—without touching it. Instead, the experimenters shoot laser beams at the sample and study what they blow loose from precisely selected regions. "We use lasers to solve chemical problems—that's the thrust of our lab," says Zare. "It's chemistry without chemistry," says Simon Clemett, a British graduate student who worked with Zare on ALH 84001, meaning that the laser techniques leave the sample untouched and thus uncontaminated—when the experimental procedures are properly followed.

"I'd gotten an interest in meteorites from my friend Peter Buseck, a chemist and geologist at Arizona State," Zare says. "He was at Stanford on sabbatical, and told me, 'I'm studying meteorites.' 'Why?' 'Because they're messengers from space.' That got my curiosity going, and I asked, 'What are you looking for in them?' and he said, 'Whatever is there.' This really got me interested in examining them, and eventually I had a reputation in the field of meteoritics." In recent years, along with his colleagues and students, Zare has been studying interplanetary dust particles, tiny meteorites that rain to Earth so softly, and so abundantly, that high-flying spy planes can collect them with sticky substances. To analyze them as precisely as the Zarelab does requires amazing attention to detail. "We spend most of our time on incredibly dull and frustrating work," maintaining high precision and checking the results against control samples, says Zare, and he ought to know, though much of the detail work on ALH 84001 was done by Clemett, whom Zare calls the "major player" in this research, and two postdoctoral fellows, Xavier Chillier and Claude Maechling.

By 1995, when Kathie Thomas-Keprta contacted Simon Clemett and asked him to analyze some samples from the Johnson Space Center, the Zarelab already knew a good deal about analyzing extraterrestrial visitors to Earth from its research into interplanetary dust particles. To keep this analysis free from unconscious bias, the three samples received from the JSC bore the

The Hunt for Life on Mars

labels "Mickey," "Minnie," and "Goofy." "That's fair," says Zare. "In the past, some people have given us blanks [samples devoid of any interesting compounds]. That's a good test of our abilities." But the samples from the JSC were, in fact, all pieces of ALH 84001.

The Zarelab was the ideal place to determine the molecular composition of a piece of ALH 84001 without contaminating the sample. The traditional techniques for removing pieces of the rock for detailed study risked adding terrestrial molecules to the bits of the rock from Mars. In that case, the most sophisticated analysis might have the perverse effect of detecting familiar compounds, not because they had been made on Mars but because they had been accidentally added in the laboratory. This would have mimicked the debacle with the Orgeuil meteorite, though in that case the contaminants entered the meteorite before it reached the laboratory.

The Zarelab's techniques avoided all direct contact with the sample. To eliminate the risk of contamination so far as possible, Zare's team placed the samples inside a vacuum chamber, where two different laser beams and an electric field were applied, initiating a three-step process that could reveal the different types of molecules at each location in each sample. The first laser puffed a cloud of matter away from the sample, the second laser selected a class of molecules within that cloud, and the electric field allowed the experimenters to count the number of each particular type of molecule in the class that the second laser had isolated. This technique has made the Zarelab a world center for certain types of chemical analysis. We can compare Zare's method to that of a detective who seeks to find out what a man's coat is made of by directing a blast of heat onto a small spot on the coat to release some of the fibers from the fabric into the atmosphere. Of these fibers, the detective arranges to blow away the strands of just one type, such as wool or silk. Then he counts the numbers of each particular size and weight of, for example, the wool fibers.

In the Zarelab, the chemists put the sample under a microscope to select the area onto which they would focus their infrared laser. Every pulse from this laser created a mini-explosion within a carefully selected region of the meteorite, which ejected material almost instantaneously. Twenty microseconds after this explosion, the team sent a beam from a second laser, this one emitting ultraviolet radiation, onto the ejected material. The ultraviolet radiation knocked electrons loose from some of the molecules in the jet, a process called ionization. By "tuning" the frequency of the second laser's ultraviolet radiation, Zare and his colleagues could select exactly which class of molecules would be ionized; other classes of molecules would be unaffected, retaining their full complement of electrons. The Zarelab then applied stage three, which would count, one by one, the individual kinds of molecules that had been ionized.

To achieve stage three's goals, the Zare team created an electric field within the vacuum chamber. The field had no effect on the nonionized particles, but it accelerated the ionized molecules to impressive velocities. Flying through the vacuum chamber at speeds approaching one hundred miles per second, the ionized molecules soon encountered a target that recorded their impacts. Their times of arrival, plus simple physics, allowed the Zarelab to discriminate among the different kinds of molecules on the basis of their weight. Each of the molecules received the same energy boost from the electric field. Just as a golf ball will move more rapidly than a football receiving the same energy input, the molecules with lower masses moved more rapidly through the chamber. "As in life, the thin ones always beat the fat ones," Zare notes. By counting the number of impacts occurring at slightly different times, the Zare team could measure the numbers of molecules of each particular mass. The mass in turn revealed the particular species within the class of molecules that the ultraviolet laser had ionized a moment before.

When scientists perform this type of experiment with a rock from Earth, they experience some difficulty in interpreting their results, simply because so many different types of molecules exist in even a small volume of most terrestrial rocks. In contrast, the parts of the Mars meteorite that lie sufficiently far from the surface to be thought free from contamination are dominated by four kinds of molecules, all of which belong to a single class. These molecules are PAHs: polycyclic aromatic hydrocarbons. As their name so clearly demonstrates, PAHs consist mainly of hydrogen and carbon atoms ("hydrocarbons") arranged into two or more adjoining hexagonal structures ("polycyclic") that chemists call benzene rings ("aromatic," as anyone who sniffs benzene can verify).

Benzene rings are one of life's most popular structural elements. Benzene itself, a colorless, highly flammable liquid, forms the simplest of the petroleum compounds. Each benzene molecule contains six carbon and six hydrogen atoms, so its chemical formula is C_6H_6.

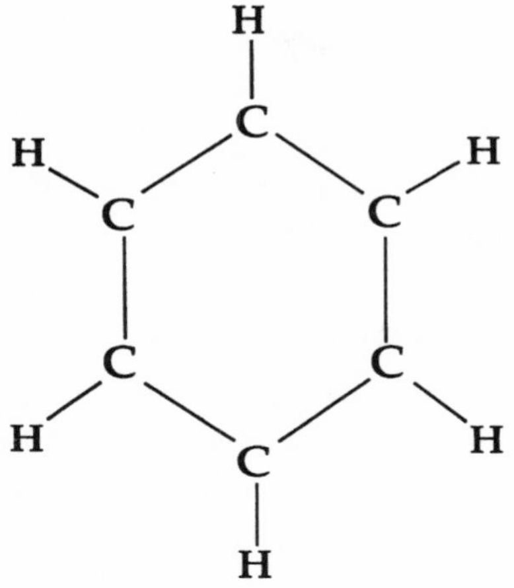

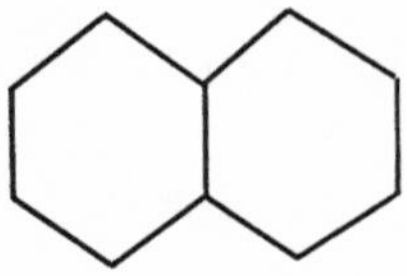

Benzene

Naphthalene

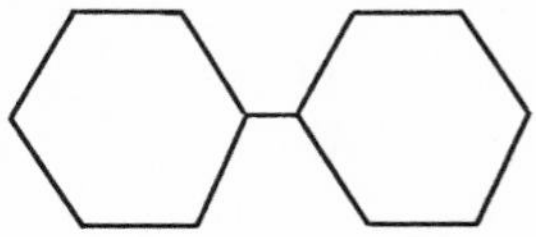

Phenylbenzene

A benzene molecule includes six carbon atoms, linked together in a ringlike structure, with six hydrogen atoms outside the ring. Molecules of polycyclic aromatic hydrocarbons (PAHs) each have two or more benzene rings, usually with additional hydrogen and carbon atoms outside the rings. For example, the simplest PAH, naphthalene, consists of two benzene rings that share a common side of their hexagons.

The six carbon atoms in benzene form a ring, with the hydrogen atoms lying outside, each of them bound to a single carbon atom. This molecular structure was first glimpsed (mentally!) by the nineteenth-century German chemist Friedrich Kekulé. After much conscious thought about the structure of benzene, Kekulé had a dream in which six snakes formed a circle, each swallowing the tail of its predecessor. In those happy pre-Freudian days, Kekulé immediately saw that his subconscious had provided an extraordinary chemical insight; his reputation was made, and few today know that a Scottish chemist named A. S. Cooper had had the same realization slightly earlier—but was late in getting it published through his contacts in France. Since Kekulé was the editor of the German journal *Annalen der Chemie*, he faced no such obstacles.

The PAHs identified in ALH 84001 contain three or four benzene rings, plus a few extra carbon and hydrogen atoms. Wherever life on Earth has existed, PAHs can be found; they are typically *not* present in living organisms themselves, but arise from the decay of living creatures, like the hydrocarbons in petroleum. "When you barbecue chicken," says Zare, "that black stuff you get on the outside if you cook it too long is just loaded with PAHs." If this were the entire story, then finding PAHs in the interior of ALH 84001 would by itself have demonstrated that life had existed on Mars and had left its decay products behind. However, this is most emphatically not the case: No one takes the presence of PAHs as proof of life. PAHs also appear in places thought to be devoid of life, such as interstellar space, old meteorites, and interplanetary dust particles.

The most intriguing aspect of the PAHs detected in ALH 84001, like that of real-estate investments, lies in their *location*. The PAHs are far more highly concentrated within the carbonate globules than outside; inside the globules, they appear in especially large amounts within the globules' black-and-white rims. "When I saw the spatial correlation between the PAHs and the carbonates, that was my 'Eureka moment,'" says Zare. "It made me say, 'I have to give [the possibility of ancient Martian life] some credence.' When that happened, I had a sleepless night, with a little bit of terror. After all, I was leading a happy life. This would obviously be controversial."

It was too late to back out, though Zare could hope to enjoy relative anonymity until the work was published. "I spoke to Kathie Thomas-Keprta, and then to David McKay and Everett Gibson," recalls Zare. "We wrote the paper by communicating with E-mail and fax, and the fax had to be secure." Early in 1996, the scientists at the Johnson Space Center, McGill University, and Stanford were ready to publish, and submitted their paper to *Science*. "I called the editor in April, and said that we were submitting an important manuscript that needed special reviewing, with no leakage," says Zare, "and he told me, 'I don't want a repetition of cold fusion.' There's no doubt that we gained a lot by going through the reviewing process, and the reviewers kept the secret," though more than one later mentioned what a difficult task this had been when they discussed extraterrestrial life with their colleagues during the spring and summer of 1996.

Then, early in August, the news leaked. "I was attending a council meeting of the National Academy of Sciences in Massachusetts," Zare remembers with no effort whatsoever. "Monday night [August 5] I flew back to Stanford, and listened on my office answering machine to an increasingly hysterical series of messages." The messages culminated in the news that Zare was expected to fly to Washington the next day in order to attend Wednesday's press conference. Owing to a series of airline problems of the sort that crop up just when one least needs them, Zare reached Washington long after midnight, feeling so woozy that he could barely present his laboratory's findings to the press. Nevertheless, referring to the possibility that meteorites could transfer organisms from one planet to another, Zare produced the most-quoted line from the press conference. "Who is to say we are not all Martians?" he asked—a subject we shall address in later chapters.

"So how do you interpret what we found?" Zare asks. "Our way [possible ancient life on Mars] seems reasonable, though amazing. It fits a lot of data, but that doesn't mean it's correct. Any scientist must be skeptical: The biggest trap in science is self-delusion, and I want to reserve the right to change my mind in the face of new interpretations or new data. When a medical doctor tells me something important, I look for a second opinion. Even the Michelson-Morley experiment [whose results support Einstein's special theory of relativity] needed confirmation before people could believe it."

Scientists agree that confirmation of the astounding results from ALH 84001 are just what we must try to obtain if we hope to draw definitive conclusions from the evidence. The teams of researchers who have looked at the carbonate globules and attempted to determine their ages, structures, compositions, and formation histories will redouble their efforts during the next few years, aware that they have a chance to prove—or disprove— what might be the discovery of the first verified extraterrestrial life. As they do so, the debate over what they have found will continue.

In November 1996, three British scientists found evidence suggesting that life may have existed in another meteorite from Mars, EETA 79001, the "Rosetta Stone" that sealed the case for the Martian origin of the SNC meteorites. Unlike ALH 84001, EETA 79001 has a relatively young age, just 180 million years, and spent only about 600,000 years in space before reaching the Earth, where it was discovered in 1979 by Antarctic meteorite hunters.

After they learned the news about ALH 84001, Colin Pillinger, Ian Wright, and Monica Grady, astronomers and meteoriticists at the Open University in Milton Keynes and the Natural History Museum in London, re-examined their samples of that meteorite and of EETA 79001. Within the carbonate globules in ALH 84001, they found compounds that apparently formed from methane, a gas that microorganisms on Earth often produce. Even more striking was their finding that as much as one part in a thousand of EETA 79001 consists of organic material. Their techniques do not yet allow them to identify the specific components of this organic matter, such as the PAHs that the Zarelab identified in ALH 84001. But the British scientists did make another fascinating discovery: The ratio of carbon-13 to carbon-12 isotope abundances in the organic matter in EETA 79001 resembles that found in living creatures and their fossils on Earth.

Why does this ratio of the numbers of carbon isotopes provide a sign of life? The enzymes that bind carbon atoms into living cells can process carbon-12 slightly more easily than carbon-13. As a result, living organisms tend to have ratios of carbon-12 to carbon-13 that are a few percent larger than the ratio in nonliving matter. In EETA 79001, the overall ratio of carbon-13 to carbon-12 equals the ratio found on Mars in general, but the carbon that forms part of the *organic* molecules in this meteorite has a noticeably different ratio, one that many consider a telltale sign of biological activity.

Even though the evidence provided by the carbon isotope ratios cannot be considered conclusive, the fact that this evidence appears in a second meteorite from Mars, and a young one at that, suggests that life on Mars may not only have existed billions of years ago, but may indeed have survived into geologically recent epochs and even into the present era. When Richard Zare learned about the British results after several months during which the results from ALH 84001 had received thorough and intense scrutiny, he sighed with relief. "So many experts have ridiculed what we have proposed that [this news] is most welcome," he said. "The findings by these well-known and well-respected meteoriticists that evidence does point to primitive life forms on ancient Mars gives to me a whole new meaning to the assertion, 'We are not alone!' "

The operations of the scientific community make it entirely reasonable for anyone who announces evidence pointing toward life on another planet to expect a wave of objections among his or her peers; some of the more spirited reactions to the news from Mars are well described by Zare's word "ridicule." To gain widespread acceptance among scientists, all conclusions, especially the most startling ones, must confront and surmount the counterarguments raised by researchers working in the same area. During the late summer and fall of 1996, a host of scientists produced critiques of the analysis and interpretation of the evidence in ALH 84001. Doing justice to these objections will require a further chapter, which we shall call "The Defense Cross-Examines."

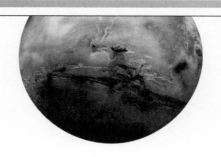

"This is

half-baked work

that should

not have

been published."

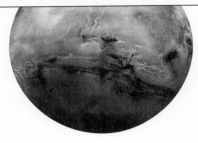

THE DEFENSE
CROSS-
EXAMINES

Suppose that we continue to employ our model of scientific debate as a courtroom drama, using it as a device to understand the pros and cons of possible life on Mars. In that case, we may imagine that Lawyer Z, prosecuting the case for ancient life on Mars to the fullest, has presented the previous chapter's discussion of the evidence. Now we may picture the defense team, led by Lawyers A and B, preparing a cross-examination intended to rebut and confound every key point that Lawyer Z alleges to have established. This rebuttal will aim to show that any conclusion that ancient life existed on Mars must be judged premature. Of course, in order to make the proceedings conform to precedent, we must then allow the prosecution an opportunity for redirect examination, followed by the defense's recross.

The Defense's Opening Statement: Do Five Possibles Make More Than a Possible?

Lawyer A, skilled in the psychology of rebuttal, will begin by congratulating David McKay, Richard Zare, and their colleagues for their clearly demonstrated skills in analyzing their sample of ALH 84001. In a magnanimous mood, he may even go out of his way to praise those who found the meteorite and demonstrated its Martian origin, providing us with the oldest rock from a planet ever found in the solar system. The defense insists, however, that *nonbiological* processes are far more likely than living organisms to have produced the evidence that the prosecution claims argues for ancient life on Mars.

The core of the prosecution's case lies in five key items that characterize this meteorite:

(1) ALH 84001 is a volcanic rock with an age of four and a half billion years whose Martian origin has been demonstrated beyond a reasonable doubt.

(2) Carbonate globules, which typically arise in the presence of liquid water, formed (at a time not well determined) within fissures in the rock, as evidenced by the shapes of some of the globules.

(3) Just at the rims of the carbonate globules, electron-microscope images reveal ovoids that resemble terrestrial microfossils, though the ovoids are smaller than any fossils found on Earth.

(4) The globules contain magnetic minerals, both iron oxides and iron sulfides, whose compositions and shapes resemble those that some bacteria produce on Earth.

(5) Organic molecules called PAHs appear preferentially within the carbonate globules; these PAHs are similar to those that result from the decay of organisms on Earth.

Before we examine these items in detail, says the chief defense attorney, let us ask a fundamental question. If, as the prosecution seems to admit, no single one of these five items *by itself* furnishes convincing evidence for ancient life on Mars, what can we gain from finding five "possibles" in ALH 84001? Does adding "possibles" together really produce a "probable," as the prosecution argued in its article for *Science* magazine? Wouldn't it be more accurate to say that if the glove might well have belonged to the defendant, but this can't be proven; if the defendant had time to commit the crime, but no witness saw him at the crime scene; if DNA evidence places the defendant at the scene, but the evidence might have been planted there; if the defendant's behavior after the crime was consistent with his guilt, but might have arisen from genuine mental confusion; and if the defendant might have been engaged in hiding evidence at his home, but no one can establish convincingly that this is so—if no single piece of evidence carries the burden of proof, how can we reasonably assert that their totality nevertheless achieves this goal?

This question has no easy answer. Like a criminal jury, each of us must decide whether the addition of possibles makes the outcome more than possible. We can agree, however, that the bedrock question remains how we judge each item of evidence before we add them together. Therefore, Lawyer A will say, let us look at the five key pieces, one after the other, and see whether we might agree that they fall short of conviction.

Nonbiological Explanations Seem at Least As Reasonable As Biological Ones

Lawyer A first attempts to dispose briskly of the five points that the prosecution adduces as evidence for life. First, everyone agrees that ALH 84001 came from Mars and has a formation age of 4.5 billion years, but this says nothing about whether the rock contains signs of life. Second, although the prosecution has not proven that the carbonate globules in the rock formed in liquid water, let us assume that they did. Even the prosecution does not argue that this provides definitive evidence for life, since carbonate globules form quite readily on Earth under natural, nonbiological circumstances. The presence of liquid water only implies an *opportunity* for life to have arisen on Mars; it would be a fine thing if we could be sure that the carbonates arose as the result of biological activity, but the prosecution must produce direct evidence that this in fact occurred.

The third item, the electron microphotographs, are replete with intriguing shapes, but we can tell nothing about the *interior* structure of these ovoids, and the identification of cell walls is one of the required, definitive signs of fossil life on Earth. The ovoids might well have been created, as such structures often are, from mineral deposits laid down without the existence of life. No paleobiologist on Earth would presume to identify objects as fossils, especially objects with these incredibly tiny sizes, on the basis of their shapes alone. Think how long it took paleobiologists to demonstrate ancient life on Earth, Lawyer A reminds the court: Three decades had to pass before 3.5-billion-year-old *terrestrial* life became generally accepted. Fourth, the presence of both iron oxide and iron sulfide in the carbonates can likewise be explained without invoking life, despite the prosecution's claim that finding them in the same place points toward their manufacture by living creatures. Finally, PAH molecules, as everyone knows or soon will know, appear throughout the universe without provoking the assertion that they are evidence of life.

The arguments that I have assigned to the defense lawyers actually come from Edward Anders, a leading authority on meteorites and the early years of the solar system, whom I call Lawyer A, and from J. William "Bill" Schopf, whom we may call Lawyer B, an expert on the earliest forms of life on Earth. In the interest of full disclosure, let me emphasize that none of the parties engaged in debating possible ancient life on Mars have cast themselves in the role of attorneys, and that I am paraphrasing their arguments. "It is wrong to maintain that I am at odds with those guys [McKay, Gibson, Zare, et al.]," says Schopf. "They just have the idea that five lines of evidence make it probable. I say it is *possible*—but it's also possible that there's no life there."

Like the final line of the McKay team's article in *Sci-*

ence, which concluded that the five items are "evidence for primitive life on early Mars," Schopf's assertion that he is not "at odds with those guys" seems a bit disingenuous. Those guys have presented the case for ancient life on Mars, quite aware of the fact that it falls short of being ironclad, while Schopf assumes the skeptical role, despite his statement at the press conference on August 7 that "I prefer to call it a discussion, not a debate." When we come to the leader of the defense team, Ed Anders, whom I have lumped together with Bill Schopf solely for convenience in assembling and comparing their arguments, we meet a man who has no hesitation in saying he *is* at odds with the authors of the *Science* article. "This is half-baked work that should not have been published," Anders says. Who is this man who minces not his opinions?

"Ed Anders was in at the beginning," says Tobias Owen, himself a leading figure in planetary science. "He is probably the leading expert on the chemistry of meteorites. It was Anders who showed that the signs of life in the Orgeuil meteorite were contamination from Earth. And, of course, it was Anders who offered to eat moon dust." During the late 1960s, before the first lunar landings, a controversy arose over the possible hazards of bringing dust and rocks from our satellite back to Earth. Despite the moon's almost complete lack of water and an atmosphere, not to mention its repeated baking and freezing at temperatures that lie far beyond the range at which terrestrial life can survive, some scientists, led by Carl Sagan, felt that we could not ignore the possibility that lunar soil might contain organisms that could harm us or our environment. When NASA announced various measures designed to quarantine the astronauts and the soil samples until they could be pronounced free of contamination, Ed Anders, driven by the basic principle that any extraterrestrial life capable of interacting with life on Earth must exist under conditions at least vaguely similar to those on our planet, wrote a letter to the *New York Times* offering to swallow some of the first lunar dust samples. As things turned out, Anders ate nothing from the moon, while the samples, which at first were kept under sterile conditions, have had a chance to infect Earth, and have not done so.

Nonbiological Carbonates, Magnetic Minerals, and PAHs

After Ed Anders had studied the article in *Science* for a few weeks, he concluded that McKay and his coworkers had overstated the case for life on Mars. As is usual within the framework of science, Anders wrote a long rebuttal, in this case to the editors of *Science* magazine; for our purposes, this letter provides a brief against the conclusion that the evidence in ALH 84001 proves that Mars harbored ancient life.

Unlike the case of the Orgeuil meteorite, in which the crucial error arose from a failure to detect *contamination* of the evidence, Anders's disagreement with McKay, Zare, et al. deals entirely with the *interpretation* of the conclusions derived from careful examination of the evidence. "The Orgeuil meteorite was bad data and bad interpretation," Anders says. "Now we have good data and bad interpretation. All that they have done is to show that the data are consistent with a biological origin. But one must consider the alternatives. You must not adopt a biological origin until you can rule out more prosaic, and therefore more plausible, explanations."

Anders's argument attempts to dispose first of the notion that the structure of the carbonate globules is most likely explained by biological processes. He calls attention to a fact that McKay and his collaborators noted: The carbonates in ALH 84001 show a variation in type from place to place within the individual globules. The inner parts of the globules contain carbonates rich in calcium and manganese, while the parts near the rims have iron- and magnesium-rich carbonates.

Anders finds an explanation for this in nonbiological processes that can form carbonates. When carbonates form in the presence of liquid water, they do so because the water carries dissolved carbon dioxide, which provides the carbonates' carbon and oxygen, and also iron, magnesium, calcium, or manganese, which combines with carbon and oxygen to make carbonate minerals. If Mars once had water with some carbon dioxide dissolved in it, the carbon dioxide would have gradually evaporated, bubbling out of the water in the same way that the carbon-dioxide bubbles in "soft drinks" do. Consider, Anders urges the jury, what happened as the carbon dioxide left the water to enter the Martian atmosphere: The acidity of the water must have changed, because carbon dioxide dissolved in water forms a weak acid. Chemists know that as the acidity of the water changes, carbonates rich in different elements—calcium, iron, magnesium, or manganese—will be preferentially deposited under different conditions of acidity. Therefore, Lawyer A can state, there is nothing surprising in finding "zoning" in the carbonate deposits, with the iron- and magnesium-rich carbonates closer to the rims of the globules than the calcium- and manganese-rich carbonates are. This is just what we expect to find in a situation with changing acidity arising from the escape of carbon dioxide from the water.

As to the presence of iron oxide and iron sulfide in the globules, Anders points out that the oldest, most primitive meteorites in the solar system, the carbonaceous chondrites, often contain similar magnetic minerals. No one seriously asserts that organisms existed in the carbonaceous chondrites (though in view of the news from ALH 84001, such assertions may yet be made). To be sure, Anders notes, the magnetite (iron oxide) in carbonaceous chondrites does not appear inside carbonate deposits, but the case for life can hardly rest on this detail. If the acidity of the water changed, we should not be surprised to find both carbonates and magnetic minerals, even if they tend to form under different acidity conditions.

What about the *shapes* of the magnetic minerals, which appear reminiscent of the magnets that bacteria produce on Earth, at least to some of those who argue for signs of ancient life in the Mars rock? In his letter to *Science*, Anders pithily notes that "such sausagelike shapes [of the minerals] are in a morphological no-man's-land." As Lawyer A would put it, we simply cannot identify the shapes as identical to those of "magnetotactic bacteria." Furthermore, no one has made careful comparisons of the shapes of the magnetic mineral grains with similar magnetic grains formed by nonbiological processes on Earth. Therefore, Lawyer A concludes, the magnetic minerals tell us nothing about the presence of ancient life on Mars.

Finally, Lawyer A may note that any attempt to explain the magnetic minerals as manufactured by bacteria on Mars must confront the fact that Mars has an extremely weak magnetic field, so weak that it has yet to be accurately measured. In the absence of Earth-like magnetism, it would seem pointless for bacteria to evolve to produce tiny magnets, if those magnets have nothing to which they can effectively respond.

Turning to the PAHs (polycyclic aromatic hydrocarbon molecules) in the carbonate globules, Anders grows no kinder, though he does praise the Zarelab for its skill in identifying them, and in establishing that ALH 84001 contains only a few varieties of PAHs. Reaching back to experiments first done in 1868, Anders points out that if you shoot methane molecules (each of which consists of four hydrogen atoms bound to a carbon atom) through a hot gun barrel, you will quickly smell naphthalene, a well-known PAH used in mothproofing compounds that forms from the reaction of methane and carbon-dioxide molecules. Modern experiments confirm that hot methane reacting on iron-rich surfaces can produce PAHs, and often yields just a few, dominant types of PAHs. Anders concludes that "for a highly evolved PAH distribution [that is, one containing only a few types of PAHs] such as that in ALH 84001, there is no way of distinguishing a biological from an abiotic [nonbiological] source."

The Defense Cross-Examines

And thus, Lawyer A will say, we have disposed of four of the prosecution's five pieces of evidence claimed to point toward ancient life on Mars. This leaves only the most visually appealing evidence, the alleged microfossils, or, since they are so small, nanofossils. Let me remind the jury, he could say, that no terrestrial fossils exist that are only 10 or 20 nanometers in diameter and 100 or 200 nanometers in length. But, he might continue, my colleague Lawyer B, who has far more experience with ancient microfossils on Earth, can present a better argument on that point.

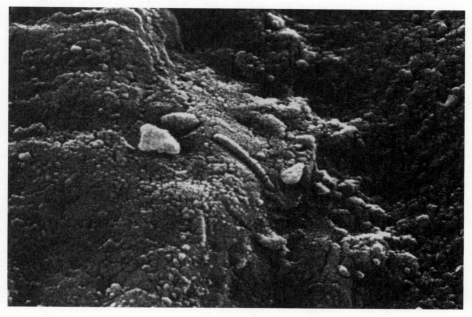

The center of this scanning electron microscope image of a piece of ALH 84001 shows a tube-shaped object that reminds some scientists of ancient fossils of life on Earth. This structure is about 500 nanometers long. (*NASA photograph*)

Interpreting the Ovoids: Nanofossils or Carbonate Excrescences?

When it comes to the search for and interpretation of ancient fossils, one of the two or three great experts is J. William Schopf, professor of paleobiology and director of the Center for the Study of Evolution and the Origin of Life at UCLA. Now in his mid-fifties, Schopf, who pronounces his name with a short *o*, studied geology at Oberlin and Harvard during the 1960s before joining the UCLA faculty, where during the past twenty-five years he has been examining the traces that life left billions of years ago in the rocks of Earth.

Schopf long held the record for the greatest age established for fossil life, 3.5 billion years; new findings have apparently raised the record age to 3.8 billion years. Schopf's fossils appear in *stromatolites* from western Australia. Stromatolites are layered rocks that form in the presence of colonies of blue-green bacteria. As sediments accumulate, they include the remnants of the bacterial corpses, which give the rock a distinctive appearance. But "I wouldn't trust stromatolites per se [as definitive evidence of life]," says Schopf, "because you can build layered [rock] structures without having bacteria present."

What *would* convince Schopf that organisms once existed within a rock billions of years old? "If you had populations of microscopic organisms whose cells had been preserved, and if you could show that the cell walls are made of organic material, and if you had populations [that is, cells of different types that show evidence of cell division] like those of the organisms we know today—then there's no doubt."

"That's like having to prove adultery," says Tobias Owen of the University of Hawaii. "You have to catch them in flagrante delicto or there's no acceptable proof." By Schopf's test, everyone agrees that the evidence in ALH 84001 does not furnish definitive proof for life.

Faced with the "nanofossils" seen in the electron-microscope photographs taken by David McKay and his collaborators, Schopf's skepticism rises high, though he loses none of his level politeness. "Those objects have diameters of about 20 nanometers—about the size of ribosomes," he notes. Ribosomes, the sites of protein synthesis in living cells, consist of protein and RNA molecules interwoven in complex fashion; they are important components of life but not life itself. "The [ovoids] are not ten times smaller but *one hundred to one thousand times smaller* than the Precambrian microfossils we know. Even small bacterial cells contain quite a bunch of ribosomes, along with mitochondria and all sorts of other things."

We shall soon meet Lawyer Z's attempt to deal with this topic, which raises the fascinating question, How small could the smallest living organisms be? "I don't know what the size limits of life are," Schopf says. "But that's not the crucial issue here. No one knows whether [the nanofossils] are fossils or not. Why not? Because no one has provided the data that can tell you. There's no data on their chemical composition: What are they made of? There's no evidence that they're made from organic matter. If they're made entirely of carbonates, then they're not life. And none of them appears to be broken, or poking out of or into the rock, as you'd expect if they had been alive. These are tiny, elongated structures. None of them seems deformed, but when we look at well-established fossils on Earth, no more than a tenth of a percent of them are sufficiently well observed to be interpretable. I call this *odd*."

Why didn't the researchers attempt to determine the chemical composition of the nanofossils? "To me it's clear why," says Schopf. "There's no shame to it: These people are geologists and mineralogists, not biologists. This is an educational problem. Physical scientists don't know biology, and vice versa." In particular, biologists would insist on slicing into the ovoid structures to see whether they are hollow, and thus more likely to have been cells that enclosed the fluids of life, as our cells do. "Technically it's a very difficult thing to do," Schopf notes. "To determine the internal structure and chemical composition of the nanofossils, laboratory experts would have to embed them in epoxy, remove them from the rest of the meteorite, cut the epoxy into thin sections with a diamond saw, and examine those sections with a transmission electron microscope." (Projects to perform these operations are now under way at the Johnson Space Center; whether they will succeed and what they will reveal remain questions for the future.)

For now, Schopf remains understandably dubious. "My rule is the one used in arms control: Trust, but verify. I believe that the observations are spot on—[the researchers] excellently reported what they saw. But can one interpret the results definitively as life? No! The glass may be half full or half empty, but no one says it's overflowing. A really relevant question is this: If these five lines of evidence had been found in a rock on Earth, would they be evidence for life? Not at all. In fact, it would be difficult to get the paper published [because the scientists who reviewed it would have called for further study]."

Redirect: The Minimum Size of Life and a Few Other Matters

During the long onslaught on his case from Lawyers A and B, Lawyer Z has maintained a stoic silence (indeed, anything else might be censured by the court), but he now rises to rebut the arguments made by the defense on cross-examination. He wisely chooses not to rehash the issue of how the jury should combine the five individual items of evidence, pointing out that the prosecution's case for life must stand or fall on how well the totality of evidence convinces those who examine it. Using time-tested legal strategy, Lawyer Z then seizes the weakest point in the opposition's case, the argument that the ovoid shapes seen at the edges of the carbonate globules are too small to have been living cells.

Note, he tells the jury, that Lawyer B frankly admits that he doesn't know how small life can be. As a result, he relies on the fact that all the life on Earth that he has studied—fossils from the early years of the planet—exists only in sizes much larger than the ovoids in ALH 84001. To Lawyer B, this demonstrates that the ovoids are unlikely to have been cells. Now, we have at least two ways to refute the assertion that the ovoids are too small. One would be to demonstrate that simple cells of the ovoids' size can exist in theory, and the other would be to show that such cells in fact exist or did exist on Earth.

As to theory, all the experts agree that what a cell requires, at a bare minimum, must be a membrane to enclose it and at least a few thousand complex molecules inside it. What we know about atoms and their chemical reactions implies that a membrane capable of performing its basic function of spatial limitation must be at least 10 nanometers thick, more or less. This means that any cell must have a thickness of 20 nanometers (to allow for a membrane on both sides) plus whatever interior space it provides—say at least 25 to 30 nanometers. Many of the ovoids appear to have nearly this thickness, and some of them are 200 nanometers long. At least one or two thousand complex molecules can fit inside a cylinder 200 nanometers long with an interior space 5 nanometers in diameter, which allows for two 10-nanometer-wide sides to a 25 nanometer-wide cylinder. One or two thousand complex molecules *may* be enough for the simplest living cell, which would then be far simpler than anything we find on Earth, where even simple cells are laden with protein-synthesizing factories and other highly developed molecular complexes. In short, we cannot rule the ovoids inadmissible as possible cells on the basis of their sizes alone.

And in fact, Lawyer Z continues, we know of bacteria on Earth, of a type called *Coxiella*, which have lengths of 200 to 400 nanometers, quite similar to those of the ovoids. To be sure, *Coxiella* bacteria are not so narrow as the ovoids, but they are considerably longer than they are wide. These are not fossils, which would indeed be more difficult to discover, but living organisms. If *Coxiella* can and do survive on Earth (so well that they are pathogens that cause diseases in humans, which explains why they have been so carefully examined despite their tiny sizes), how can we dismiss the ovoids as possible ancient cells? We also know that bacteria exist one to two miles below the surface in the Columbia River basalts in Washington, completely out of touch with the surface environment. When conditions grow particularly harsh, many of these bacteria can lower their metabolism and shrink from their normal sizes, which are a few microns across, to just one-tenth of those sizes in their attempts to deal with starvation. These "dwarf bacteria" remain barely alive by reducing their rate of cell division to once every century or even longer.

While I am at it, says Lawyer Z, forgetting that he offered to confine his remarks to the single topic of the minimum size for life, let me point out to the distinguished counsel for the defense that we are discussing not the possibility of life on Mars now, but rather whether life existed on Mars billions of years ago. Although Mars *today* has an extremely weak magnetic field, all experts agree that Mars could easily have lost an original magnetic field as the planet cooled and its presumably iron-rich core solidified. Quite honestly, we do not know how strong that field was many billion years ago, and our understanding of how planets generate and maintain magnetic fields is as uncertain as our knowledge of how life began. But we cannot dismiss the argument about magnetotactic bacteria on the basis of an unknown component, the strength of the magnetic field at the time that these ancient carbonates were formed. Let me also point out, Lawyer Z could say, that when we find nonbiological magnetite on Earth, as in lava flows, some titanium is usually mixed in with the magnetite. The bacteria that make magnetite do not mix it with titanium, nor do we find titanium mixed with the magnetic minerals in ALH 84001.

And if I may discuss just one more point, Lawyer Z might say while his junior colleagues try to tug him into his seat before he utterly exhausts the attention of the jury, let me deal with my distinguished colleague's attempt to dismiss the evidence in the PAHs as just another example of what we find throughout the cosmos. The PAHs, which even the defense concedes are not the result of contamination either in the laboratory or while the meteorite was in the ice, are *not* like the PAHs found in other meteorites, the sort that have never been to Mars. The distribution of PAH types in the carbonate globules is *in*consistent with the types found in ordinary meteorites—but *is* consistent with the types that appear when bacteria decompose. Furthermore, consider the *location* of the PAHs, which are concentrated in the carbonate globules, and indeed within certain parts of the globules. Even Lawyer A, the outstanding advocate on a nonliving origin for these molecules, agrees that he knows no nonbiological process that would cause this to occur.

I could go on, Lawyer Z would add, but I risk taxing your patience. Let me close by suggesting that when you have five items of evidence, each of which is consistent with ancient life on Mars, they *do* provide a much stronger argument for life than each one by itself. Of course, each of you must judge for yourself, but I urge you to find for life on Mars.

Recross: *The Temperature at Which the Carbonate Globules Formed*

Rising for his final chance to present evidence, Lawyer A brings the jury back to the question of how the all-important carbonate globules formed within the meteorite from Mars. Brushing aside the prosecution's objection that new evidence cannot now be introduced, he refers the jury to the studies of the carbonates by Ralph Harvey and Harry McSween, Jr., geologists at Case Western Reserve University and the University of Tennessee, respectively, who are experts in the interpretation of the minerals found in meteorites.

One month before the news from ALH 84001 broke, Harvey and McSween published an article in *Nature* that analyzed the composition of the meteorite, which was already of great interest as the oldest rock from Mars. In direct contradiction to the work by Chris Romanek and his collaborators, they concluded that the carbonates formed not at temperatures between 0 to 80 Celsius but instead at temperatures above 650 Celsius (1,200 degrees Fahrenheit). Harvey and McSween propose that the carbonates originated when a huge impact on Mars caused silica-rich rock, similar in composition to the basalts that form from volcanically erupted material, to meet with fluids that carried significant amounts of carbon dioxide. Under the right circumstances, such an encounter could produce carbonates with bands of color similar to those of the globules in ALH 84001. If the carbonates arose in this manner, Lawyer A will argue, no one would believe that they have anything to do with life. We shall pass over Lawyer Z's passionate attempts to rebut this argument with measurements of the numbers of different isotopes, which some scientists claim can reveal the temperature at which the carbonates formed. The arguments are complex, relatively tedious, and do not convince those who, like Lawyer A, can marshal counterarguments.

In closing, Lawyer A might say to the jury, let me urge you to pay close attention to the prosecution's arguments for life. Lawyer Z asserts that the types of PAHs found in ALH 84001 are *consistent* with the decay of bacteria. But has he shown that these are the precise types of PAHs that *do* form when bacteria decay? No, he has not. And do not be fooled by the prosecution's favorite phrase—"consistent with life." Remember what Sherlock Holmes said: You must eliminate the other possibilities before adopting an improbable conclusion. In order to conclude that ancient life existed on Mars, you must demand and receive proof that all nonbiological explanations *must* be rejected.

How Wide Is the Gulf Between the Prosecution and the Defense?

Sparing the reader the final arguments for the prosecution and the defense, which would summarize the opposing viewpoints in terms of "adds up to life" and "cannot disprove the alternative," we may ask whether the entire structure that we have created does not amount to a straw person: Do the scientists who consider the evidence from ALH 84001 really disagree to the point that they can be accurately depicted as advocates for and against a conclusion of ancient life on Mars?

For validation of their conclusions, scientists do not appeal to a jury of representative citizens, but only to each other, or, more precisely, to those who are familiar with the subject area under discussion. Scientists aim not to win the case but to produce evidence and hypotheses that will bear close scrutiny by experts who know that an important part of their job is to approach all claims skeptically. On another level, however, scientists do function as advocates. Every scientist succeeds or fails to the extent that his or her results receive validation, and this requires, in all but the most exceptional cases, a willingness to argue on behalf of one's own ideas. These arguments take place under well-understood rules of engagement, but they are indeed arguments, not rarefied symposia with no consequences for the participants.

Even cursory acquaintance with the key players in the debate over ancient life on Mars reveals a wide variety of personalities and differing shades of willingness to engage in argument. The spectrum extends at least from Ed Anders, at the feisty end, to the careful statements of Bill Schopf, whom I have capriciously teamed with Anders on the defense team. Likewise, the prosecution team includes various styles of scientific discussion. Not one of them asserts that the evidence proves that life once existed on Mars, yet it is easy to discern that many of them feel that it did, and that ALH 84001 offers supporting evidence. On the other hand, many scientists feel certain that the meteorite from Mars contains no convincing evidence for life, and some find the situation so overblown as to verge on the ridiculous. All the scientists agree that further investigations, especially on Mars itself, will reveal the answers with far greater clarity than we can now achieve.

Any geologist will be happy to say that if we look at a typical two- or three-billion-year-old Earth rock, we are highly unlikely to find any signs of life in it, yet we know that life was flourishing all over our planet at that time. And these are sedimentary rocks, which form under water and within which fossils slowly form. No igneous (volcanic) rock on Earth has fossil evidence for life—not one. Therefore, finding convincing evidence for ancient life on Mars in an igneous rock that is 4.5 billion years old and contains carbonates that formed billions of years ago would indeed be an improbable event. The statements in this paragraph will, I feel sure, receive agreement from all of the scientists mentioned in this book. All that remains is to determine whether the improbable in fact has occurred.

To assess the probability for ancient life on Mars, we should devote some attention to its closest relative, ancient life on Earth. The next chapter deals with the questions of how life began and what conditions may be generally required for life to have originated. If we had the answers to these questions, we would be much closer to knowing the likelihood of life, both past and present, on Mars or any other planet. Even the modest amount that we *do* know about the origin of life provides valuable insights into the chances for ancient life on Mars.

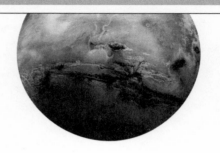

"There's a

bazillion

hypotheses.

There are

many scenarios."

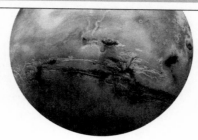

THE ORIGIN
OF LIFE

T he discovery of life beyond Earth would be an astounding event, provoking wonder and awe among most of humanity. Beyond its emotional impact, finding life on another planet would allow us to take a giant step toward the resolution of a crucial issue regarding life in the cosmos: How closely does extraterrestrial life resemble life on Earth?

The answer to this question should eventually reveal whether similar conditions on different worlds have produced similar types of life. Any resemblances might have originated from "cosmic seeding"—the direct transfer of living cells from one world to another—or from the fact that evolution often converges to results that appear similar because they represent adaptation to similar situations. The first discovery of extraterrestrial life, no matter how primitive that life may be, will immediately provide a basis for comparing life-forms on two different worlds. With any luck, this comparison will quickly reveal whether the cosmic-seeding hypothesis may be valid, and whether our concepts of how life originates and evolves stand the test of newfound reality. For now, if we hope to speculate intelligently about life on another planet, we must begin by examining the one form of life that we know—the one here on Earth—to draw conclusions that may apply to life in general.

The Importance of DNA to Earth Life

The totality of all life on Earth, for which I shall introduce the abbreviation "Earth life," depends on a long-chain molecule called DNA (deoxyribonucleic acid). Every DNA molecule consists of two spirals, woven around each other and joined by thousands of cross-links, pairs of molecules that lie between the spirals. Each of the cross-linking molecules, called a "nucleotide," fits with just one other type of nucleotide. *The sequence in which the cross-linking nucleotides appear in DNA specifies how an organism behaves.* Therefore, every organism needs a way to maintain the information coded in its DNA, and to preserve this information when its cells make copies of themselves. Our own cells, to pick a convenient example, continually reproduce themselves, though some do so far more quickly than others.

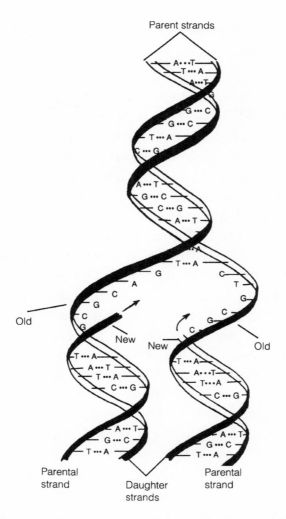

Parent strands

Old

New   New

Old

Parental
strand

Daughter
strands

Parental
strand

DNA molecules each consist of two long spirals made of sugar and phosphate molecules, linked by pairs of molecules called nucleotides. To make the two strands link together properly, the nucleotide adenine (A) can pair only with thymine (T), while guanine (G) can pair only with cytosine (C). During replication, the DNA molecule divides lengthwise, and each of its two strands then reconstitutes a "daughter strand" from molecules floating within the cell. Because of the requirements for a correct "fit" of the two strands, the two "daughter" DNA molecules are identical to the "parent molecule."

The Hunt for Life on Mars

When cells reproduce, every DNA molecule in the cell splits lengthwise. This separates each of the thousands of cross-linking pairs of molecules, creating two individual spirals, each with a series of partial cross-links jutting out from it. Floating in a bath of molecules, each of the spirals can reconstitute its missing partner from raw materials floating within the cell by adding the missing members of the cross-linking pairs, along with a second spiral backbone. In this process, called replication, each half of the cross-linking pairs builds a new partner just like its old one, not by accident but because only a molecule identical to the former link can fit properly between the old and new spirals. The rebuilding of the cross-links resembles the actions of a sightless person who can reconstitute pairs of socks from hosts of individuals, relying on touch alone to sense that only a certain type of sock will match the one at hand, while all others in the drawer do not. The replication process produces two identical DNA molecules where only one existed before, and each of these "daughter molecules" can itself replicate, with the processes continuing as many times as the organism requires, producing additional versions of the original.

DNA molecules govern the entire functioning of living cells, and their replication lies at the heart of biological reproduction. The details of our DNA molecules to a large extent determine who we are and how we are, as well as what our descendants will be like. Each person's DNA molecules are unique, as identifiable as a fingerprint and used like fingerprints in criminal detective work. If a perpetrator leaves some of his or her cells (these are typically blood cells if violent activity occurs) at the scene of a crime, DNA experts can determine—with a chance of error so small as to be negligible—whether or not the DNA in those cells matches the DNA in the cells of a suspect. Great criminal cases have turned on such seeming arcana.

Mutations and Evolution

During the process of DNA replication, small changes called *mutations* can occur, both at random and as the result of outside influences (for example, by exposure to radioactivity or certain toxic chemicals). The result of a mutation is that one of the daughter molecules differs slightly from its parent. Most mutations produce results that are neutral or even harmful to the organism in its attempts to survive and to reproduce; they therefore quickly disappear from the scene. However, some mutational changes tell the organism to do something additional that proves useful in its quest to survive and to reproduce. In that case, provided that the mutation can be passed from ancestors to descendants, the organisms carrying the mutation may come to dominate the local scene, and can eventually produce new types of organisms.

The process of evolution rests on the competition for reproductive success; that is, on competition within a species, not between different species, as might seem the case at first glance. In this competition, organisms with more descendants win the right to express themselves genetically at the expense of their fellow species members. Charles Darwin was the first to see how competition plus variation (the result of mutations) could lead to new species. Today, with Darwin's insights repeatedly validated, biologists also know how the evolutionary process works in terms of DNA molecules.

The Genome Writes the Book of Life

In every DNA molecule, different parts of the sequence of cross-linking nucleotides specify how to make particular types of protein molecules, which play a significant role in how the organism functions. A sequence that specifies the production of a particular protein molecule is called a *gene*. The order of all the cross-linking pairs in an organism's DNA molecules therefore provides the organism's *genome:* the full set of all its genes. The genome writes the "book of life," the story of how any individual organism grows and functions.

To write this book of life, nature billions of years ago evolved the *genetic code*, in which each letter consists of three successive cross-linking nucleotides in DNA, which specify a particular amino acid from the set of 20 that Earth life uses. To describe a protein typically requires 100 to 500 cross-links—something like a hundred letters in the genetic code. Thus, taken a hundred or so at a time, the letters in the genetic code form the "sentences" that we call genes, and the organism's full complement of

THE GENETIC CODE

Each Triplet of Nucleotides -- U, C, A, or G -- Specifies an Amino Acid

Second Letter

		U	C	A	G	
U	UUU } phenyl-alanine UUC UUA } leucine UUG	UCU UCC } serine UCA UCG	UAU } tyrosine UAC UAA stop UAG stop	UGU } cysteine UGC UGA stop UGG tryptophan	U C A G
C	CUA CUC } leucine CUA CUG	CCU CCC } proline CCA CCG	CAU } histidine CAC CAA } glutamine CAG	CGU CGC } arginine CGA CGG	U C A G
A	AUU AUC } isoleucine AUA AUG methionine or start	ACU ACC } threonine ACA ACG	AAU } asparagine AAC AAA } lysine AAG	AGU } serine AGC AGA } arginine AGG	U C A G
G	GUU GUC } valine GUA GUG	GCU GCC } alanine GCA GCG	GAU } aspartic acid GAC GAA } glutamic acid GAG	GGU GGC } glycine GGA GGG	U C A G

The "genetic code" describes the fact that each sequence of three nucleotides specifies a particular amino acid, or else the start or stop of a particular gene sequence. The code shown here refers to RNA molecules, which carry the information embodied in the genetic code from place to place within living cells; these molecules employ uracil (denoted U) in the same way that DNA molecules use thymine.

sentences writes its genome. Biologists have understood the genetic code for decades, and have begun to read the book of life for many different types of organisms. The "human genome project," an attempt to read the entire genome that describes human beings, should reach completion during the first decade of the coming millennium.

The Unity of Earth Life

From the mightiest redwood to the smallest microbe, Earth life—the totality of all living creatures on our planet—consists of one basic type of life. This is not a statement of the simple truth that all this life exists on or near the Earth's surface. When scientists say that Earth has only one form of life, they mean that *all life on Earth depends on the same types of chemical reactions among the same types of molecules*. For example, proteins, which are complex molecules widely used by living organisms, are each made from 20 different small and simple molecules, the amino acids. Although literally thousands of amino acids could exist, and do exist in laboratory experiments, Earth life has evolved to use just these 20, and never more, in every type of protein found in any organism. This implies that interactions among amino-acid molecules must have been of crucial importance for the origin of Earth life.

When it comes to replication, every form of life on Earth employs the same DNA molecules. The details of how the cross-links are ordered along the spiral determine one type of life from another, but the fundamental structure applies to all Earth life: From whales to the Columbia River bacteria two miles underground, the same types of molecules form the two spiral backbones, and the same types of other molecules provide the cross-links connecting the spirals. All Earth life speaks with an identical language, using the same genetic code to express its instructions, without a change of accent or dialect. The planetary oneness of DNA and the genetic code it embodies speaks eloquently of the unity of Earth life. To understand this unity, we must determine how life on Earth originated, and how it developed this planet-wide language. Until we find extraterrestrial forms of life, all the clues to this determination lie in the most ancient forms of Earth life that we can find.

What Do We Know About the Earliest Life on Earth?

Though the origins of Earth life remain shrouded in mystery, we have steadily increased our understanding of its early development and evolution. Since an organism's genes control its development, any evolutionary changes must have been accompanied—in fact, mandated—by a change in its genes. Hence biologists measure the similarity of organisms by checking the resemblance between the sequences of their genes. To do so, they usually examine not the DNA itself but a closely related molecule called RNA (ribonucleic acid).

RNA molecules, which resemble single strands of DNA, copy the genetic information carried in DNA molecules and assist in making proteins. One kind of RNA (messenger RNA) carries genetic information to the sites where proteins are made, while other kinds (transfer and ribosomal RNA) help to make the protein molecules. All types of RNA read the information carried in DNA molecules, which furnish a "master copy" of the organism's gene sequences. During the past two decades, biologists have examined the RNA molecules from thousands of different types of organisms. In some cases, they have managed to read the entire genome, the full sequence of genes that specifies everything about that species.

To compare different types of organisms whose genes have been well described, biologists measure the difference between their genes, which provides the *evolutionary distance* between any two types of organisms. The measurements of these evolutionary distances made during the past two decades have produced a revolution in our understanding of how Earth life developed. This revolution began with the work of Carl Woese of the University of Illinois, who devoted years of studies to the RNA of different microorganisms, and eventually overthrew the dominant paradigm of how life evolved. Woese published his chief insights during the mid-1970s, but they have only recently received widespread, though not yet total, acceptance among evolutionary biologists.

Before Woese's work, biologists classified all Earth life into one of two types: *prokaryotes* and *eukaryotes*. Prokaryotes (Greek for "before the nucleus") consist of single cells that have no well-defined *nucleus*—no specific part of the cell holding the cell's DNA molecules. In contrast, eukaryotes ("good nucleus") may be single-celled or multicellular but all their cells have well-defined nuclei: They each contain a region bounded by a membrane that encloses the cell's DNA. This division allows eukaryotic cells to concentrate their DNA, and to include far more of it—typically ten to a thousand times more—in each cell than a prokaryotic cell does. In addition to well-defined nuclei, eukaryotic cells typically include other well-defined entities, called *organelles* ("little organs"), that perform specialized functions within the cell. Bacteria are all prokaryotes, while eukaryotes include animals, plants, fungi, slime molds, and a host of single-celled organisms. As eukaryotes ourselves, we have a prejudice in their favor, and respond more readily to complex, multicellular organisms. Introductory biology courses usually skip quickly past prokaryotes and devote perhaps twenty times more attention to eukaryotes. Yet bacteria—prokaryotes—have been the dominant and most successful form of Earth life for more than three billion years.

Common sense suggests that prokaryotes, being so much simpler than eukaryotes, must have evolved first. Furthermore, an entirely reasonable hypothesis proposes that eukaryotes evolved from prokaryotes by incorporating several different types of prokaryotes into a single eukaryotic cell. Unfortunately for this conclusion, however, Carl Woese showed that a third basic type of cell exists. Woese recognized that some of the organisms classified as bacteria—the prototypical prokaryotes—have RNA that differs so much from the RNA in bacteria that they have no business being lumped in with the prokaryotes. Instead, these organisms deserve to be classified as a third type of Earth life, which Woese named Archaeabacteria ("ancient bacteria") but which are now called Archaea to avoid confusion with bacteria. Like prokaryotes, Archaea lack a well-defined nucleus in each cell, but they are so different from prokaryotes—basically in terms of their RNA sequences—that they have been elevated to a third distinct "superkingdom" of life.

Woese's work has led most biologists to conclude that Earth life, like Caesar's ancient Gaul, divides into three *domains:* Archaea, bacteria, and eukaryotes. Measurements of the evolutionary distances between repre-

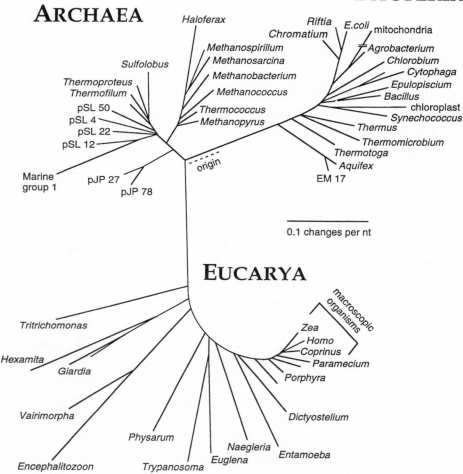

This diagram shows the "three domains" of Earth life: Archaea, Bacteria, and Eucarya. The distances between different organisms shown in the diagram represent evolutionary distances, based on the similarities of the organisms' DNA and RNA. Note that most of the subject matter of a conventional biology course, organisms larger than a few hundred microns in size, occupy only the extreme lower right of the diagram. (*Courtesy of Norman Pace*)

sentatives of these three domains have led to a surprising conclusion: Eukaryotes are at least as closely related to the Archaea as they are to bacteria! Far from having evolved from what were formerly called prokaryotes, eukaryotes have at least as ancient an origin. In fact, it seems likely that bacteria—the former prokaryotes—will prove to be a branch from the Archaea–eukaryote lineage. The presence of organelles in most eukaryotic cells may be a red herring that misled biologists until they learned how to read the genetic information within living cells. Billions of years after the first eukaryotes appeared, eukaryotic cells may well have incorporated prokaryotic cells, but the origin of eukaryotes would still lie at least as far back in time as the start of the bacteria.

In the summer of 1996, the Institute for Genomic Research in Rockville, Maryland, an organization seeking to map the entire human genome, announced the first complete sequencing of the genome of one of the Archaea, an organism called *Methanococcus jannaschii*, which we may abbreviate as "M.j." M.j.'s genome contains 1,738 genes—about two percent of the total number estimated to exist in the human genome. Of these 1,738 genes, more than a thousand are unlike any genes in prokaryotes or eukaryotes. The remaining genes show that M.j. indeed has a closer genetic resemblance to eukaryotes than to prokaryotes.

How do these M.j.'s make a living? Like many of the Archaea, they are hyperthermophiles, organisms that love extreme heat. M.j.'s live near deep-sea vents called "white smokers"—locations that until recently remained completely uninvestigated by humans. In 1982, the deep-sea scientific submarine *Alvin*, one of the mechanical heroes of biological research, made a dive two miles into the Pacific Ocean near the Galápagos Islands, where hot water gushes through vents in the sea floor, surrounded by clusters of strange wormlike creatures and enormous mats consisting of tiny organisms. In the material that *Alvin* brought to the surface, biologists found a new organism, the now-familiar M.j., which thrives at temperatures between 120 and 205 degrees Fahrenheit (50 to 90 degrees Celsius), and at pressures two hundred times the surface atmospheric pressure we enjoy.

M.j. cannot stand oxygen. Instead, it requires carbon dioxide, nitrogen, and hydrogen, all of which spew from the white smokers it calls home. Instead of relying on solar energy, M.j. draws its energy supply from chemical reactions between hydrogen and carbon dioxide. Some of these reactions produce methane gas: The "methano" contained in M.j.'s first name refers to the fact that M.j. produces methane as a by-product of its metabolism, just as we exhale carbon dioxide and plants produce oxygen. Since methane closely resembles the natural gas we consume as fuel, M.j.'s metabolic processes deserve closer study. This strange organism, and the Archaea that resemble it, may eventually show us better ways to make natural gas by using organic, nonpolluting pathways.

As their name implies, the Archaea appear to rank among the oldest forms of Earth life. Furthermore, when biologists plot the evolutionary distances between different types of Archaea and bacteria, they find the closest resemblances between Archaea and bacteria that are both thermophilic—lovers of high temperatures. Most Archaea are thermophiles, but most bacteria are not. Though we should hesitate to leap prematurely (look what happened to the theory that eukaryotes evolved from pro-karyotes!), the conclusion begs to be drawn: The earliest forms of life on Earth appeared near deep-sea vents, where hot water, carrying all sorts of simple molecules, gushed from the sea floor.

This conclusion becomes even more attractive when we note that the early Earth—our planet during the first half-billion or billion years after its formation—was the site of volcanic activity far greater than occurs now. The existence of Archaea shows that deep-sea vents can easily provide the raw materials for life, which carry an energy supply with them in the form of potential chemical reactions between hydrogen and carbon-dioxide molecules. This fact in turn shows that sunlight, which we tend to regard as a requirement for Earth life, should be regarded as a highly useful, but not essential, source of energy.

When Did Life Begin on Earth?

For a century and a half, biologists have attempted to find and to date our original ancestors. During the 1870s, scientists aboard the British vessel *Challenger* dragged samples of muck from the ocean bottoms, rich in fantastic bottom-dwelling sea creatures but devoid of the primordial ooze that some of the scientists had hoped to find, the *Urschleim* (original slime) thought to have given rise to all forms of Earth life. Today biologists recognize that the search for *Urschleim* was doomed to failure, because the planet has changed so much since the time life began that no original life-forms could endure in the oceans—or anywhere else, for that matter. To name the most evident change, Earth life has "polluted" the atmosphere with large amounts of oxygen, produced by hosts of blue-green bacteria floating in the oceans. This "pollution," which occurred about two billion years ago, eliminated most types of anaerobic (oxygen-avoiding) bacteria, though some have found places of refuge— for example, in our stomachs, where oxygen does not penetrate. Evolution has made Earth quite a different planet from what it was when life began, four billion years ago or more.

How can we date the origin of life? The basic method consists of finding rocks with fossils, and using radiometric (radioactive-decay) techniques to place an age on the rocks. Provided that we can be sure that the fossils formed along with the rocks, this method works well. This is the case for sedimentary rocks, which formed underwater from slowly deposited layers of debris, but as we saw in the case of ALH 84001, igneous (volcanic) rocks pose a greater challenge, since we can be sure only that whatever worked its way into the rock did so *after* the rock had formed.

Paleobiologists, who study fossil forms of life, have now pushed the dates of the oldest known forms of life back to at least 3.5 billion years ago. However, these forms of life hardly rank as the first on Earth. To begin with, we know them as fossils, which means that they must have developed structures sufficiently pronounced to be recognized billions of years later. In addition, these fossils consist of cells strung together in chains. Almost certainly, earlier forms of life consisted of individual,

The Origin of Life

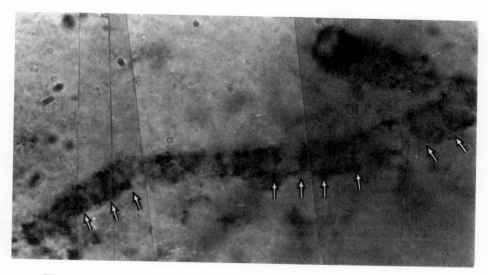

This photograph shows the oldest certified fossils of Earth life, found in rocks in western Australia by J. William Schopf and dated to a time 3.5 billion years ago. Note that this organism apparently already had many cells, some of which are indicated by the arrows. This implies that life on Earth originated well over 3.5 billion years ago. (*Courtesy of J. William Schopf*)

free-floating cells, and before that of chemical mixtures only loosely confined within some sort of precellular structure. The oldest rocks found on Earth, from the Isua formation in Greenland, date back to an era 3.8 billion years before now. For decades, paleobiologists have debated whether or not the Isua rocks contain convincing evidence of life without reaching any firm conclusion. The news from ALH 84001 resonates with this debate in a new light, since the carbonates in the Mars meteorite, difficult though they are to date, may have ages comparable to the 3.8 billion years of the Isua rocks and the 3.5 billion years of the oldest confirmed fossils of Earth life.

What Earth Was Like When Life Began

What, then, were the conditions on the primitive Earth, the "Wonder-bread years" between 3.8 and 4.5 billion years ago, when Earth life began its existence and evolution? Because we possess no Earth rocks older than 3.8 billion years, our conclusions must rest on speculation about the formation of the solar system, extrapolation from eras closer to the present time, and information from the oldest rocks in the solar system: the meteorites (all about 4.5 billion years old), the oldest rocks found on the lunar surface (about 4.2 billion years), and ALH 84001, the single ancient rock from Mars.

This approach produces a picture of the primitive Earth quite different from the planet we know today. Today the Earth has relatively well-defined continents and seas; in those days, the Earth constantly thrust volcanic material up through its seas, only to see the mid-ocean ridges destroyed by the subsidence of the crust beneath them, which had not stiffened to the degree we find today. In those eras, mountain-sized chunks of cosmic debris struck the Earth every few centuries, and much larger objects impacted many times in each million years. These bombardments eventually used up most of the debris in interplanetary space, so that impacts from large objects have become much rarer. Today we have a fairly stable atmosphere (human activities aside!) formed mainly from nitrogen and oxygen; the primitive Earth's atmosphere consisted mainly of carbon dioxide and nitrogen, which underwent significant changes from the impacting objects. The end of the era of intense bombardment, at a time dated between 3.8 billion and 4.0 billion

The Origin of Life

years before the present, marked the final stages of the processes that built the Earth and the other planets and that had already left them with their basic sizes and shapes 4.5 billion years ago.

The largest impacts on the primitive Earth—and on primitive Mars as well—might have blown a significant fraction of the then-existing atmosphere into interplanetary space. However, comets colliding with Earth and Mars also brought *new* supplies of relatively light compounds, such as ice and frozen carbon dioxide, to the planets they struck. On both planets, whether or not life began and then survived may have depended on whether the final key impacts did more to bring in or blast away the atmosphere and other volatiles near the planets' surfaces.

Human exploration of the moon, which allows astronomers and geologists to assign ages to many of the moon's features, provides us with some confidence in setting a time 3.8 billion years ago as the end of the era of intense bombardment. This time marks the end of the era when the basins containing the moon's now-frozen lava plains and most of its craters formed; since then, only occasional large impacts have occurred. By extrapolating, we can apply the same history to Earth and Mars. This extrapolation seems reasonable because conditions should have been roughly the same throughout the entire inner solar system, from Mercury's orbit out past Mars's, during the time that the solar system formed and the last great rains of debris struck the newly made planets. As the solar system's first 700 million years drew to a close, the era of intense bombardment ended, because almost all the debris left over from the formation of the solar system either had been exhausted through impact or had achieved stable orbits that did not intersect the planets' paths around the sun. Throughout the 3.8 billion years that followed, impacts have been a rarity, though large ones, such as the dinosaur-killer 65 million years ago, have played an important role in the evolution of life on Earth by clearing ecological niches for the survivors of these impacts.

In the violent and changeable circumstances on the primitive Earth, life may have begun not once but many times, only to be destroyed when a particularly large object struck the Earth. This must almost certainly have been true if life needs only a relatively short time—a few tens of millions of years, let us say—to arise under favorable circumstances. The 700 million years on the primitive Earth would have included dozens of these time periods. Likewise, if the impacts transferred life-bearing rocks from Earth to Mars, or from Mars to Earth, the arrival of these rocks from another planet might well have occurred not once but dozens, hundreds, or even thousands of times, depending on how many rocks each impact blasted loose. Even if most of these transfers failed to strike a spark on the new planet, the large number of interplanetary rock voyages suggests that primitive Earth may well have received samples of early Martian life, if it existed, and that Mars should have received a smaller, but still significant number, of life-bearing rocks from Earth. On the other hand, the life forms in these rocks would probably lose a competition with any homegrown organisms, which would have arisen and evolved in response to the local conditions.

The Origin of Earth Life

Now that we have identified the three types of Earth life and have assigned a date earlier than 3.8 billion years ago to their common origin, we may ask, as many have before us, How did life begin on Earth? The plain fact is that we do not know. "There's a bazillion hypotheses," says John Baross, a microbiologist at the University of Washington and an expert on the oldest forms of life. "There are many scenarios." The answer to the origin of life may eventually prove to be that life arrived on Earth—perhaps encapsulated within a meteorite—and that it already had a genome something like that of the Archaea. Or Archaea could be close to the original form in which life itself arose on Earth, as near to *Urschleim* as we can hope to come.

One of our chief areas of ignorance in determining the origin of Earth life lies in its location: We do not know whether life began in the deep oceans, where the Archaea now lie, or at the boundaries between water and land, or in still other environments. But didn't high school biology teach us that life began in warm ponds and tidal pools? Like the "fact" that eukaryotes evolved from prokaryotes, the tidal-pool theory may prove untenable, thanks to a better understanding of conditions on the early Earth.

The most obvious characteristic of Earth's surface today—its division into oceans and land masses—quite likely did not exist early in the Earth's history, because the Earth's crust had not yet grown sufficiently stiff to support the continents. Instead of continents, we should picture an entire planet something like today's mid-

Atlantic ridge, where volcanic material seeps upward from below, slowly spreading the crust and moving Europe and Africa farther from the Americas. "Think of a thick volcanic sludge, rich in organic tar," says Norman Pace, a microbiologist at the University of California who worked with Carl Woese. "That's a perfect setting for chemical reactions on the surfaces of mineral grains that could lead to life." In Pace's view, every "hot spot" in the ocean floor could have been a site where life originated. Somehow, quite early in its history, Earth life produced prokaryotes and eukaryotes in addition to the Archaea. In those bygone eras, when the Earth lacked the continental stability that it now possesses, the only truly stable environments would have been those in the deep ocean.

The hypothesis that life originated in volcanic sludge, rather than within "some warm little pond," as Charles Darwin imagined, must answer a question that the tidal-pond theory can perhaps satisfy more easily. How do small molecules link into larger ones, and in particular, how do they form the long-chain molecules characteristic of Earth life, each of which contains hundreds or thousands of smaller molecules, repeating themselves over and over? The edges and bottoms of ponds and tidal pools offer excellent places for large molecules to organize themselves into long chains. These basins undergo repetitive cycles of drying out and refilling that tends to organize molecules into chain-like structures. Deep-ocean sludge pots also offer places where water runs over a solid surface and encounters numerous small grains where chemical reactions occur more easily, but they lack the cycles of evaporation and rewetting that helps to organize long-chain molecules at the edges of ponds and pools each time that the water dries out.

How Long Does Life Take to Appear on a Favorable Planet?

Earth's history is mainly that of a planet with life, whose record extends over at least 3.5 billion years of the 4.5-billion-year total. Since the fossils from 3.5 billion years ago already show a fair amount of evolutionary development, we can state with some certainty that if Earth offers a representative example, life requires less than half a billion years to originate in a favorable environment. But could life appear far more rapidly than this, say in a hundred million years? Or ten million? The latter number still allows the possibility of multiple originations and multiple exterminations of life on Earth, and by implication on other planets as well. Since complex molecules tend to be fragile, "Use them or lose them" may be the general rule about the chances that they will lead to living creatures. It could be that unless life appears on a planet within a few thousand years, or perhaps a few million, after conditions become generally favorable, it will not arise at all unless the conditions change significantly.

From the evolutionary distances between different members of the Archaea, we do know that the earliest forms of Earth life evolved only slowly, because Archaea dated hundreds of millions of years apart differ only modestly. The slow pace of early evolution seems reasonable, since the steady appearance of more varied forms of life should have increased the competition among organisms that drives evolution, and thus the rate of evolution itself. As we noted above, the Archaea teach another key lesson about the earliest forms of Earth life. They were all thermophiles, lovers of high temperature. "Early life was high-temperature and slowly evolving," Norman Pace likes to say. "And there's another old myth that has vanished. Early life was autotrophic, capable of getting its own 'food' and not dependent on finding food floating in the environment. Of course, the 'food' was basically hydrogen and carbon-dioxide molecules, and the organisms made methane out of them, just as the same sort of methanogens do today."

The fact that none of the Archaea were known to science a few decades ago teaches us that we should not be overly quick to believe that we possess all the information needed to unravel the mystery of life's origin. Fortunately, we have another arrow in the quiver in addition to finding and studying the oldest forms of Earth life: We can attempt to make life ourselves.

Modeling the Origin of Life: The Miller-Urey Experiment

Neither the tidal-pool nor the ocean-sludge hypothesis can yet answer all the questions involved in explaining how life began. When biologists discuss the origin of life, the conversation often turns to scientists' best effort at making life in a test tube: the Miller-Urey experiment. In 1953, only a stone's throw from the football stadium at the University of Chicago where Enrico Fermi had overseen the first human-controlled nuclear chain reaction a decade earlier, Harold Urey, a Nobel prize–winning chemist, worked with a graduate student named Stanley Miller to re-create "prebiotic" conditions on Earth. Miller had suggested this unconventional experiment, and had managed to overcome Urey's objection that nothing might come of it—leaving Miller years away from his Ph.D.—by agreeing to devote less than a year to it.

Attempting to reproduce the conditions existing on Earth four billion years ago, Miller and Urey built a closed system consisting of two flasks, an upper and a lower one, linked by two glass tubes. They partly filled the lower flask with water, to model the oceans, and above the water injected a gaseous mixture of methane, ammonia, hydrogen, and water vapor, to duplicate the essence of the Earth's primitive atmosphere. Miller and Urey then heated the water, producing steam that drove some of the gaseous mixture through one of the tubes into the upper flask. There the gas received a jolt of energy in the form of an electrical discharge, similar to lightning discharges on Earth. In the final step of the experiment, some of the gases from the upper flask passed downward through the second tube, which led them through a condenser and so back to the lower flask. Thus the Miller-Urey experiment reproduced the cycle of evaporation and rainfall on Earth, with an energy input to the mixture during its gaseous phase.

The only objections to the experiment came from doubting chemists, one of whom asked Urey at a seminar, "What do you expect to get?" "Beilstein!" replied Urey, referring to the author of the classic *Handbuch der Organischen Chemie*, which began as a single volume but now runs through many hundred, all of them devoted to the properties of the molecular compounds classified as "organic"—that is, based on carbon molecules as a key structural element.

Miller and Urey watched their experiment run for a few days. What did it produce? Not all the compounds in Beilstein's book, but a sludge of many different molecules, among which, in relatively large numbers, were . . . amino acids! Amino acids, the "building blocks" of Earth life, are relatively small molecules, containing 13 to 27 atoms, that link together to form much larger protein molecules, each consisting of several hundred amino-acid molecules. The atoms in amino-acid molecules are mainly hydrogen, carbon, oxygen, and nitrogen—the most abundant types of atoms in the universe, not counting helium and neon (which do not bind into molecules), and which were available in plenty on Earth.

Why are proteins so important? Not counting water, proteins provide more than half the mass of living cells. Within these cells, different kinds of proteins perform an amazing variety of tasks, including structural support, energy storage, signaling, movement, defense against strange substances, and speeding up certain types of chemical reactions (we call these particular proteins *enzymes*). Proteins rank among the most complex molecules known, with widely varying shapes and sizes; a human body works by employing tens of thousands of different types of protein molecules, each with a specialized function. Think of a child playing with an enormous

The Origin of Life

set of blocks that come in just 20 different types, with each block a letter of the alphabet. If the child strings together "sentences" that are hundreds of letters long, most of the sentences will be nonsense, as indeed almost all random strings of amino acids are in terms of their usefulness to living cells. But a few, highly complex permutations will be immensely important. If the child happens to spell out "There is no more basic information about biology at the molecular level than the fact that just twenty amino acids link together to form all types of proteins," he would achieve something comparable to the complexity that a living creature embodies by forming a single large protein molecule from available amino acids.

When you learn that the Miller-Urey experiment produced amino acids but failed to yield any protein molecules, you can react in at least two ways. You can be astounded, as Miller and Urey were, that the experiment yielded significant numbers of amino acids, or you can reflect that the model of primitive Earth took the process of making complex molecules only as far as amino acids, none of which contains more than 11 carbon atoms, or 15 hydrogen atoms, or 4 oxygen or nitrogen atoms. That's a long way from the complexity found in the simplest Earth life—but is it so far from the minimum complexity that life requires?

As we have seen, the analysis of the ovoids in ALH 84001 raised the question, What minimum number of molecules can produce life? Experiments like the one that Miller and Urey performed may eventually provide the answer to this question.

Amino Acids in Meteorites

Just as the production of amino acids in the Miller-Urey experiment did not amount to making life in a test tube, so too the discovery of amino acids in a meteorite, startling though it may appear to be, does not prove that life exists in interplanetary space.

On September 28, 1969, a meteorite fell near Murchison, Australia, and was soon discovered to belong to the most primitive class of carbonaceous chondrites, and thus a representative of the first chunks of matter to coalesce as the solar system began to form, just over four and a half billion years ago. Studies of the compounds inside the meteorite revealed the presence of 74 different types of amino acids, as well as all five of the small molecules that make the cross-links in DNA and RNA. The possibility of contamination was ruled out, because the meteorite was newly found and the amino acids lay well inside it; a few years later, amino acids were also found in another meteorite, which fell near Murray, Kentucky, in 1950.

However, it seems quite certain that the amino acids in the Murchison and Murray meteorites did *not* arise in living systems. This conclusion rests on the fact that most types of amino acids can appear in one of two varieties, mirror images of each other, like a left-handed and a right-handed glove. Life on Earth has chosen just one of these two varieties, the left-handed one, and contains none of the other possibility, the right-handed amino acids. This has occurred because chemical reactions proceed more efficiently when molecules of just one "handedness" are present. As life began, one type of handedness presumably had a slight advantage in numbers over the other, simply by chance. Because the more abundant type was favored to produce more molecules like itself, simple competition led to its complete dominance in all terrestrial life forms.

Biologists expect any life that appears in the universe to employ only right-handed or left-handed amino acids or their functional equivalents. In contrast, nonbiological processes should produce equal numbers of left- and right-handed molecules, as we find in the Murchison and Murray meteorites. The handedness of the amino acids therefore provides a key test for life. To a "prosecutor" who argues for life on the grounds that amino acids are present, the "defense" can reply: If the glove doesn't fit, you must acquit. The existence of equal mixtures of the two handednesses proves a nonbiological origin beyond a reasonable doubt, whereas a predominance of left- over right-handedness, or the inverse, furnishes a strong indication that biological selection has been at work. David McKay and his collaborators are now engaged in testing ALH 84001 to see whether it contains any amino acids. Stay alert to the news, and remember to ask, if anyone tells you that amino acids exist in the meteorite from Mars, What sort of handedness do those molecules have?

How Well Did the Miller-Urey Experiment Model the Early Earth?

Astronomers' concepts of the primitive Earth now differ markedly from what was believed four decades ago. (Furthermore, the Mars results imply that we should not be too hasty to conclude that Earth life arose here; it might have had its origin under conditions quite different from those on the primitive Earth and made the long journey through interplanetary or even interstellar space.) So far as the conditions on the primitive Earth are concerned, the major change in astronomers' conclusions deals with the amount of hydrogen in Earth's primitive atmosphere. Instead of the atmosphere rich in hydrogen atoms that Miller and Urey modeled in their experiment, most research now points toward a relatively hydrogen-poor atmosphere, with far more carbon monoxide and carbon dioxide. Experiments made with hydrogen-poor model atmospheres have yielded many of the molecules used in life, although they have not yielded the amino acids that the Miller-Urey experiment so stunningly produced. These later experiments do produce plenty of hydrogen cyanide, which can produce some types of amino acids when it floats in liquids resembling the early oceans and is subjected to ultraviolet radiation. Since Earth's primitive atmosphere did not prevent solar ultraviolet radiation from reaching the surface, we find conditions on the primitive Earth, according to the most up-to-date models, that were at least *favorable* to making amino acids.

One important lesson of the Miller-Urey experiment, which was run through several variants, resides in its demonstration that no matter what the atmospheric composition may have been, *none of this works unless oxygen was absent from the primitive atmosphere.* Oxygen prevents the formation of amino acids, or even their precursors, in a mixture of methane, hydrogen, ammonia, and water vapor: It was poison to life even before life began. Because our own existence depends on oxygen, we may have trouble imagining it as poison, but the property that makes oxygen useful when you can deal with it—its eagerness to combine with many simple atoms and molecules, the process called *oxidation*—makes it big trouble for any chemical system that will be disrupted when oxygen atoms grab hold of, and combine with, the molecules in the system. If we think of oxidation as slow burning or extremely rapid rusting (which in fact it is), we can recognize that oxidation will destroy any organism not prepared to deal with it. Two billion years ago, when microscopic organisms began to release large quantities of oxygen into our atmosphere, other forms of life exposed to the "pollution" had to either hide from it, adapt to it, or perish. Humans and all other animals resulted from a positive adaptation to the rise of oxygen in the atmosphere of Earth and evolved to use the oxygen, rather than to react to it as poisonous.

The successful origination, persistence, radiation, and evolution of Earth life leads immediately to the questions, What conditions are required for similar processes to produce life elsewhere in the universe? And How often do these conditions occur? At the risk of sounding an old refrain, we must bring out the twin rejoinders, "We don't know the answers now" and "Find other forms of life and we'll know far more."

Panspermia

We should not neglect a third possibility for life's origins—that life arrived from other worlds through cosmic seeding.

The Hunt for Life on Mars

In 1903, soon after Percival Lowell's observations of Martian canals had raised popular interest in life on Mars to new heights, a Swedish chemist named Svante Arrhenius wrote an article for the German magazine *Die Umschau* suggesting that life, encapsulated in spores, could travel through interplanetary and even interstellar space, developing into a new array of living organisms if the spores happened to find a hospitable environment. "According to this view," Arrhenius wrote, "it is quite conceivable that the living beings on all planets are related, and that a planet, as soon as it can shelter organic life, is soon occupied by such organic life." Arrhenius recognized that if this were true, discerning the origin of life would become even more difficult than determining how life arose on the planet under direct observation. Arrhenius's concept of cosmic seeding came to be called *panspermia*. On the shelf for many years, cosmic seeding leapt back into prominence when the news from ALH 84001 broke, provoking Richard Zare's suggestion that we could all be Martians. For that matter, the Martians could be Earthlings, if a rock from our planet containing living organisms reached Mars during its early years and seeded it with Earth life.

Could life survive an interplanetary journey lasting millions of years, encased in a rock such as ALH 84001? Surprisingly, the answer may well be yes. Some terrestrial bacteria seem capable of spending indefinite time on "hold," dried out and completely nonfunctional, but still entirely capable of resuming their bacterial duties once they are reexposed to water and warmth. (Of course, experiments along these lines so far cover only a few years, not thousands or millions.) Ultraviolet radiation in space presents a health hazard, but even a fraction of an inch of a rock's outer layer provides complete protection from ultraviolet to any organisms inside a meteoroid larger than a pebble. Cosmic rays—fast-moving particles that pass through all of interplanetary space—likewise can damage or destroy even dormant bacteria, but they cannot penetrate more than a few inches into rock.

Any good-sized meteoroid could therefore easily protect life in its interior. The greatest hazard on an interplanetary journey would probably arise from radioactivity *inside* the rock. An aggregate of matter similar in composition to a meteoroid will include small numbers of radioactive nuclei, including radioactive isotopes of uranium, thorium, rubidium, and potassium. The decay of these nuclei, which are distributed throughout the object, produces fast-moving particles, many of them electrons, that could eventually kill any organisms. However, even if the average meteoroid took millions of years to travel from planet to planet, as ALH 84001 did, some lucky ones, blasted into just the right sort of trajectories, could make the trip in only thousands of years. In that case, dormant organisms inside them could survive the effects of the radioactive nuclei.

Directed Panspermia

The discovery and analysis of Martian life would reveal a great deal about whether cosmic seeds have passed between the sun's third and fourth planets. While waiting for this happy event, we may consider whether cosmic seeding might occur not by haphazard chance but as the deliberate attempt of life on one planet to seed another. Seventy years after Arrhenius's first publication, two scientists at the Salk Institute in San Diego, Francis Crick and Leslie Orgel, the former a Nobel prize–winning biologist who helped discover the structure of DNA and the latter equally respected, proposed a theory of "directed panspermia," according to which extraterrestrial societies might have chosen deliberately to "infect" other planets with their forms of life. In this effort, microorganisms would be far more cost-productive than larger forms of life. Crick and Orgel envisioned an interstellar spaceship with perhaps a ton of microorganisms—a hundred samples of each of a thousand trillion different bacteria. Traveling at modest velocities, the spaceship could infect most of the planets in the Milky Way within a few hundred million years of its launch.

Crick and Orgel wondered what reason we, or some other species, might have for polluting other planets, and thought that we might find a motive to do so if we concluded that we are alone in the cosmos and decide to enrich its attributes by diffusing life. Other scientists have found a possible motive in the "cosmic zoo" hypothesis, which proposes that Earth life may be the result of another civilization's biological experiment. Like good scientists, Crick and Orgel asked whether any evidence supports the notion that other civilizations sent microorganisms that started life on Earth, and found two arguments for this hypothesis. One is the fact that all Earth life writes its book of life with the same genetic code, instead of with several different, perhaps similar ones, which one might expect if life began at various sites on Earth at different times. The second is that many organisms on Earth require not only carbon, hydrogen, oxygen, and nitrogen, the basic types of atoms found in all Earth life, but also atoms much rarer on Earth, such as chromium, nickel, and molybdenum. If—and so far this if has not become anything more than hypothesis—we found planets where these atoms are far more abundant than on our planet, this might be a sort of cosmic fingerprint revealing the place that sent life to Earth. Everything we have learned about the cosmos, however, points in the opposite direction: The relative amounts of different types of atoms are nearly the same in star after star, with some exceptions that do not change the basic pattern.

Directed cosmic seeding, therefore, remains only a tantalizing idea, to be tested, like cosmic seeding in general, by discovering forms of life beyond the Earth. "I see three basic ways that life could have appeared on Earth," says Leslie Orgel. "First, it might have been brought here by comets or meteorites. Second, the Miller–Urey experiment [modeling tide pools] might be the way to go. Third, life could have begun in the deep-sea vents." For now, these three possibilities for bringing forth life on Earth all remain viable. If we can find extraterrestrial life, and can determine both its resemblance to Earth life and the approximate date of its origin, we should be able to conclude far more about the ease with which life appears in the cosmos. The location where we find life should reveal far more about the method most favored for the origin of life. Although the meteorite from Mars seems to bring this knowledge closer, we are still awaiting its arrival.

With this thought, we can return to the planet that dominates our thinking about extraterrestrial life, the fourth rock from the sun: Mars.

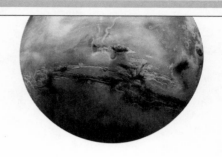

"This seat

of Mars,

This other

Eden,

demi-paradise"

CAN LIFE EXIST ON MARS NOW?

Throughout the twentieth century, humanity has rightly regarded Mars as the extraterrestrial planet most likely to have life. Of the sun's nine planets, only two—Earth and Mars—have been washed by water and possess atmospheres that are at least vaguely supportive of the existence of life. Mars has a rotation period almost identical to Earth's: Each Martian day equals just over $24^{1}/_{2}$ hours. Even the tilt of Mars's rotation axis resembles Earth's (24 degrees instead of $23^{1}/_{2}$), so that Mars also has seasonal variations in temperature and amounts of sunshine that are similar to those on Earth.

Unlike Earth, however, Mars cannot have liquid water on its surface today, and its thin atmosphere contains mainly carbon dioxide, with only the barest traces of water vapor. The polar caps of Mars likewise consist mostly of frozen carbon dioxide, along with small amounts of ordinary ice. To explain the similarities and differences between Earth and Mars, and to assess the chance that life exists on Mars now, we must perform a threefold examination: of the processes that formed these planets four and a half billion years ago, of the planets' subsequent evolution, and finally of the results from the miniaturized laboratories that we sent to Mars to search for life two decades ago.

Fourth Rock from the Sun: Does It Have the Wrong Location for Life?

Two basic facts—the planets' sizes and distances from the sun—explain most of the differences between Earth and Mars. Considerably smaller than Earth, Mars has only 54 percent of Earth's diameter and just 11 percent of Earth's mass. Because of its low mass, Mars exerts only 40 percent of Earth's gravitational force on objects at its surface. Noticeably farther from the sun than Earth, Mars maintains a distance from the sun that varies between 130 and 160 million miles, while the Earth, moving along a more circular path, remains between 91 and 94 million miles from the sun. As a result of its greater distance, and because the intensity of sunlight decreases in proportion to the square of an object's distance from the sun, each square foot on Mars receives less than *half* of the radiant solar energy falling on a square foot on Earth. The combination of its low mass and low solar heating leaves Mars with an uphill fight to maintain the conditions that favor the origin and continuation of life.

Both of these handicaps to life are linked to Mars's location in the solar system. Mars's greater distance from the sun produces the lower amount of solar heating, while its low mass arose from the fact that Mars had the ill fortune to form relatively close to Jupiter, the sun's fifth and by far most massive planet. During the era of planet formation, at times between about 4.3 billion and 4.6 billion years ago, the entire solar system—the sun, its planets and their satellites, plus a host of asteroids, comets, and meteoroids—condensed from a rotating cloud of gas and dust. For reasons still largely unknown, this cloud began to contract under the influence of its own gravitation. As it contracted, the cloud flattened and rotated more rapidly, eventually becoming a flat pancake of matter larger than the present solar system. Within this rotating pancake, collisions among the particles of dust produced millions of mountain-sized clumps of matter, which astronomers call "planetesimals." Each planetesimal moved in orbit around the central, densest part of the cloud, which was destined to become the sun. Encounters among the planetesimals then produced large aggregates: the planets, their large satellites, and the largest asteroids.

The Struggle for Mass Among Planetesimals

Like a family of young birds clamoring for food, each planetesimal in effect sought to maximize its mass at the expense of its neighbors, using gravity as the tool for acquiring matter. The biblical maxim "He that hath, to him shall be given" ruled this competition, awarding runaway success to a relatively few objects. Although each of the planetesimals exerted gravitational force on all the others, any close encounter between a larger and a smaller object tended to add the smaller one's mass to that of the larger planetesimal. These captures increased the larger object's mass, giving it a still better chance of attracting other planetesimals.

Eventually, while the bulk of the original matter joined the sun, most of the mass in the planetesimals coalesced to form one of the sun's planets or one of their larger satellites. Jupiter, the sun's most massive planet, has 318 times the Earth's mass—but only $1/1,000$ the mass of the sun. Saturn has 95 Earth masses; Uranus has 15; Neptune has 17. Jupiter and Saturn consist largely of hydrogen and helium, which also provide 99 percent of the sun's mass. Because they formed at distances ranging from 5 to 30 times Earth's distance from the sun, the four giant planets could retain large amounts of hydrogen and helium, the two lightest gases. The sun's heat, which tended to evaporate the lightest gas, had a lesser effect on regions farther from the sun. Since these more distant regions included greater volumes of space, they tended to include more matter, which in turn exerted greater gravitational forces. In contrast, the regions closer to the sun could never "grow" planetesimals sufficiently massive to retain hydrogen and helium. Once the sun began to shine, its warmth evaporated almost all of these two elements from the inner solar system. As a result, the sun's four inner planets—Mercury, Venus,

Can Life Exist on Mars Now?

Earth, and Mars—consist mainly of silicon, oxygen, aluminum, and iron, with almost no hydrogen and helium. All the hydrogen trapped in the oceans of Earth, whose average depth equals less than $1/1{,}000$ of the Earth's radius, provide nowhere near one percent of the total mass of our planet.

Earth did best among the inner planets in the competition for material. Venus, the Earth's near-twin in size and mass, has 81 percent of Earth's mass; Mercury, even smaller than Mars, has a mere 5.5 percent. As a result of its low mass and its proximity to the sun, Mercury cannot prevent gas from escaping and has essentially no atmosphere at all. Venus, in contrast, has an atmosphere of carbon dioxide so thick that it weighs on the surface with a pressure almost one hundred times greater than the surface pressure on Earth. This atmosphere is just what we expect to develop on an Earth-sized planet, if living organisms do not form carbonate rocks that lock up most of the carbon dioxide near the surface, preventing the existence of a thick carbon-dioxide atmosphere.

Like Venus, Mars does have a carbon-dioxide atmosphere, but one so thin that it provides a surface pressure less than $1/100$ of the pressure on Earth. In this low pressure, the result of Mars's low mass, lies the explanation for the total absence of liquid water from the Martian surface.

Why Can't Liquid Water Exist on Mars Now?

Liquid water cannot exist on the Martian surface because water on Mars behaves the way that carbon dioxide does on Earth. The atmospheric pressure at the Martian surface equals only 6 millibars—0.6 percent of the pressure at Earth's surface. This amount falls below the minimum pressure that allows any water to exist as a liquid. Instead, any ice whose temperature rises above 32 degrees Fahrenheit (0 Celsius) promptly sublimates into water vapor, and any water vapor that cools below the freezing point will condense directly into solid ice.

Anyone who has used boiling water to cook an egg in a high-altitude city such as Denver or Quito knows that water boils there at lower temperatures than it does at sea level. This occurs for the same reason that some people experience difficulty breathing at high altitudes: The atmospheric pressure decreases as the altitude increases. It is the atmospheric pressure that prevents water from boiling until its molecules begin to acquire speeds sufficient to let them escape from the liquid and join the atmosphere. At sea level, boiling occurs at a temperature of 212 degrees Fahrenheit (100 Celsius), but at higher altitudes, with a lower atmospheric pressure, the boiling temperature decreases. At the top of Mount Whitney or the Matterhorn, three miles above sea level, the pressure falls to half of its sea-level value, and the boiling point equals just 175 F (80 C). We can imagine ascending still higher in the atmosphere, watching the boiling point fall to 120 F, to 80 F, and finally all the way to 32 F, the freezing point of water. At this point, the atmospheric pressure will equal only $1/165$—0.6 percent—of its sea-level value. By coincidence, this pressure almost exactly equals the average surface pressure on Mars today.

When the boiling point falls to the freezing point, no

liquid water can exist. Instead, any ice that warms to 32 degrees Fahrenheit will immediately turn to water vapor, but never to liquid water. This would duplicate what we observe for carbon dioxide on Earth. Since carbon dioxide remains frozen at temperatures less than -71 degrees Fahrenheit (-57 Celsius), ice-cream vendors use chunks of this "dry ice" to keep their wares cold. When the dry ice warms, it sublimates quickly into carbon-dioxide gas as soon as it reaches -71 F. The name "dry ice" rightly implies that no liquid carbon dioxide appears. If we want to produce liquid carbon dioxide, we must raise the pressure to 5.1 atmospheres—5.1 times the sea-level atmospheric pressure, and equal to the pressure that a diver feels 125 feet deep in the ocean. At this pressure, carbon dioxide will turn to liquid once it warms to -71 F, and the higher we raise the pressure, the higher the temperature must rise before the liquid evaporates. With 70 atmospheres of pressure, we can maintain liquid carbon dioxide up to temperatures of 85 degrees Fahrenheit (30 C).

No Water, No Life

In the absence of liquid water, we have no expectation of life. Everything we know about the conditions under which Earth life can survive leads us to conclude that the presence of liquid water is a requirement for life. We can make two exceptions to this general rule. First, we can imagine that another substance, such as ammonia or methyl alcohol, plays the role of a solvent for extraterrestrial life, allowing molecules to float and interact within it, and buffering environmental change so that a rapid rise in the temperature, for example, will not cause an organism to cease functioning. However, water appears to offer potential forms of life not only the best solvent, but also the solvent made most easily from the most abundant atoms in the cosmos. Second, some forms of life on Earth do not require liquid water at all stages of their existence, only during the crucial ones. Nevertheless, we can remain firm in our insistence that until proven wrong, we expect to find life only where we find at least some liquid water.

Does this mean that no life can exist on Mars? Not at all! The low atmospheric pressure disqualifies water from existing as a liquid on the Martian *surface*. Possibilities for finding water on Mars therefore include the edges of the polar caps, where pockets of liquid water might form under the ice; underground caverns, which might have some residual heating from the days when volcanoes were active on Mars; and the entire subsurface of Mars, where all indications point to sizable amounts of water permanently frozen into the soil. This possibility of permafrost means that Mars's subsurface may resemble the Siberian tundra, where the permafrost extends thousands of feet underground and contains colonies of bacteria that thrive under the local conditions, which allow tiny amounts of water to be liquid at any moment.

The bacteria in the Siberian permafrost almost certainly evolved from ancestors that needed more liquid water and gradually adapted to the harsh conditions of frozen soil. We should not conclude that their existence implies that Earth has *no* environment too harsh for life.

Can Life Exist on Mars Now?

Detectable Earth life appears to be entirely absent from two large portions of our planet: the Antarctic ice sheet below its uppermost layers, and newly formed mountains of lava created in places such as the Hawaiian Islands. The latter gradually become colonized by life from nearby regions, but the vast mass of ice in Antarctica remains a biological vacuum.

The hopes for finding life on Mars now seem to rest on discovering places where water can exist as a liquid despite the fact that we see no water on the Martian surface and can demonstrate that we expect none. On the other hand, the surface of Mars contains abundant evidence that liquid water *has* existed in copious amounts, leaving behind lake beds, channels, and even modest floodplains. In short, Mars has changed over the years from a planet with liquid water on its surface to one without.

Long-term Changes in Planetary Conditions

Far from enduring in a single, permanent state, at least two of the sun's planets have undergone impressive planet-wide changes over the 4.5 billion years since they formed. Earth has changed the most, thanks to the evolution of life on and near its surface, and to the plate-tectonic activity that continuously moves and buries pieces of its crust. Life on Earth has locked up most of the carbon near its surface, leaving us in a relatively carbon-starved situation. If this were not so, carbon in the form of carbon dioxide would dominate our atmosphere, leaving us sweltering within a stifling supergreenhouse produced by carbon dioxide's ability to absorb heat. In addition, Earth life has enriched our air in oxygen, so that instead of an atmosphere rich in carbon dioxide with a nitrogen admixture, we have a thinner atmosphere, four-fifths nitrogen and one-fifth oxygen. On Mars we see (as yet!) no signs that life has caused similar changes to the planet. Yet Mars too has undergone significant alteration, changing from a pleasant to a hostile planet, so far as life is concerned. The causes of these changes remain obscure, though astronomers and geologists do not lack ideas on the subject. We do know that Mars once had rivers and lakes, running water, evaporation, and rainfall. Much of the evidence supporting this conclusion existed before detailed examination of ALH 84001 added weight to the evidence for an old, wet Mars.

The Evidence for Liquid Water on Mars

Because Mars was born small, with only half the Earth's diameter and one-tenth of Earth's mass, it fell short of Earth's ability to capture and retain volatile molecules such as nitrogen, H_2O, and carbon dioxide. These are the compounds that gave Earth and Mars their atmospheres. Mars was born comparatively poor in volatiles, and grew still poorer as it aged. Four billion years ago, Mars had liquid water on its surface, but as the planet grew older, its water gradually disappeared.

How do we know that Mars had liquid water in bygone eras? The proof came two and a half decades ago, when the *Mariner* spacecraft photographed the planet's entire surface. These photographs, confirmed in more detail by the *Viking* orbiters, show sinuous channels, which

Can Life Exist on Mars Now?

geologists feel sure must have been carved by liquid in motion, as well as dried-up lake beds and at least one example of a sudden flow of water on a large scale. The only liquid that is likely ever to have existed on Mars is water, whose molecules are made from two of the most abundant elements in the solar system. Spectroscopic measurements of the sunlight reflected from the Martian polar caps show that they consist mainly of frozen carbon dioxide, but a fraction of a percent of their total volume consists of water ice, and a tiny amount of water vapor exists in the planet's thin atmosphere.

Mars almost certainly never had planet-wide oceans, as Earth did and to a large extent still does. Four billion years ago, however, Mars had liquid water coursing across its surface and possibly raining from its skies. The atmosphere must have been much thicker then, capable of providing both the larger greenhouse effect and the higher surface pressure needed to allow liquid water to exist on Mars.

This atmosphere presumably arose from volcanic activity. Today the Martian volcanoes are dormant or extinct, though their enormous bulk makes them the tallest mountains in the solar system, mute evidence that Mars's surface once cracked and moaned with volcanic activity that released hydrogen, carbon monoxide, carbon dioxide, and nitrogen gases trapped in the planet's subsurface layers. Even greater amounts of these gases should have arrived from space trapped in icy planetesimals and comets, during the era of intense bombardment that ended 3.8 billion years ago.

As a result of this bombardment and the release of gases by volcanoes, billions of years ago Mars should

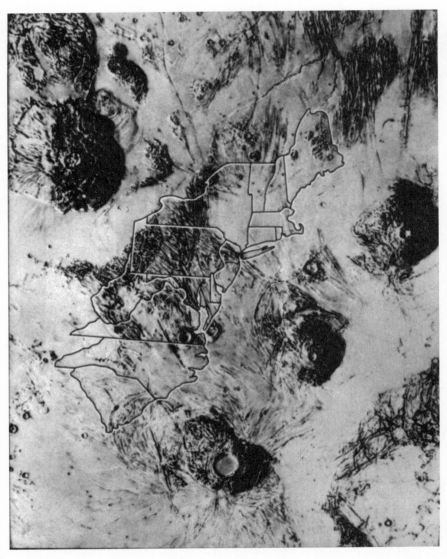

Mars's four great extinct volcanoes appear in this photograph, with the eastern United States superimposed to show the scale of sizes. The highest of the volcanoes, Olympus Mons, 400 miles wide and 15 miles high, is the largest mountain in the solar system. (*NASA photograph*)

have had a thick atmosphere made primarily of two gases: carbon dioxide, which forms the bulk of the present atmosphere of Mars, and water vapor. If these two substances had an abundance sufficient to increase the current atmospheric pressure by a factor of ten, water could have existed as a liquid within the temperature range from 32 to 70 degrees Fahrenheit (0 to 21 degrees Celsius). If instead the Martian atmosphere contained one hundred times the amount of carbon dioxide that it does now, the atmospheric pressure would have risen to 60 percent of the surface pressure on Earth, extending the temperature range for liquid water to span the interval from 32 to 185 degrees Fahrenheit (0 to 85 degrees Celsius). Planetary experts think that the primitive Martian atmosphere was still thicker—perhaps as thick as Earth's is now. If the additional atmospheric pressure arose not only by carbon dioxide but also by water vapor in the atmosphere, so much the better: Water's ability to remain liquid depends only on the total atmospheric pressure, and a larger amount of water vapor makes a cycle of evaporation and rainfall more likely.

Mars: That Other Eden, Demi-Paradise

Thus primitive Mars—the planet as it was more than four billion years ago, when it was less than half a billion years old—had abundant liquid water and an atmosphere dozens or hundreds of times thicker than it is now. Although scientists do not concur as to whether complete cycles of rainfall, evaporation, and further rainfall occurred on primitive Mars, they do agree that liquid water carved the channels, still visible four billion years later, and once filled some of the craters whose interiors show ancient shorelines. Primitive Mars was not only wetter than Mars today, but warmer as well. The greater amounts of atmospheric carbon dioxide and water vapor would have increased the "greenhouse effect," the ability of the atmosphere to trap infrared radiation from the surface, and thus to keep the planet warm. Even today, points on the Martian equator reach noontime temperatures of 60 or 70 degrees Fahrenheit (15 to 21 Celsius), though they cool to more than 100 degrees below zero by midnight, so poorly does the thin atmosphere retain heat. With an atmosphere ten times thicker than the current one, Mars could have had slightly higher daytime temperatures and much warmer temperatures at night.

In Shakespeare's *King Richard II*, the monarch's uncle, John of Gaunt, speaks of "[t]his earth of majesty, this seat of Mars/This other Eden, demi-paradise." Even though Gaunt's "seat of Mars" was in fact a reference to England, the characterization of Mars as "this other Eden, demi-paradise" serves well to describe the scene on the Martian surface, four billion years ago, when rain fell, water flowed, volcanoes belched, and the conditions for life were all in place.

Where Did the Water Go?

What happened to this demi-paradise of primitive Mars? Somehow, the atmosphere steadily grew thinner. Some

These wide, winding channels on Mars seem to have been carved by water in motion. The numbers of small craters visible in the beds of the channels allow these channels to be dated back at least 3 billion years, based on what astronomers know about the rate at which meteorites have bombarded Mars. (*NASA photograph*)

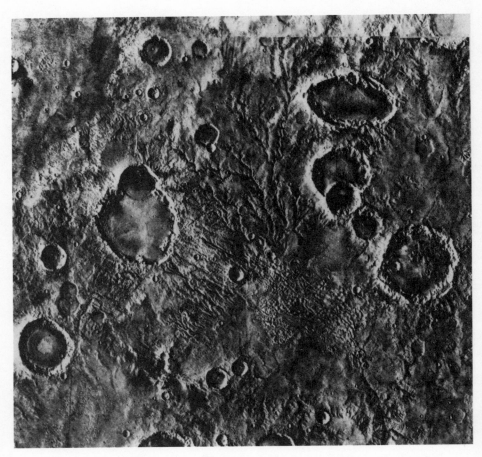

This region on Mars, called Margaritifer Sinus, includes what seems to be an ancient lake basin. The complex of thin channels visible near the center of the photograph suggests that water existed for prolonged times near this lake. Mars experts consider this an excellent place to search for sedimentary rocks that might preserve fossil life. (*NASA photograph*)

Can Life Exist on Mars Now?

This large triangular valley, more than 20 miles wide, may have arisen from a sudden collapse of the Martian surface, perhaps when permafrost or ice below ground suddenly melted, producing a gigantic flash flood that carved the channel that extends to the left. (*NASA photograph*)

of the carbon dioxide may have escaped into space, and a small fraction of the atmospheric CO_2 froze into the Martian polar caps. Some of the water vapor likewise escaped, while a tiny amount joined the dry ice of the polar caps; and the bulk of the water, quite possibly, remains widely distributed over the planet, not *on* but *under* the Martian surface.

Wherever the carbon dioxide and water vapor may have gone, the fact they disappeared almost totally from

The Hunt for Life on Mars

Mars's atmosphere has had a sad effect on the Martian environment. As the atmospheric pressure decreased, so too did the amount of the atmospheric greenhouse effect, because the atmosphere contained fewer molecules that prevented heat from escaping from the planet's surface. As the greenhouse effect decreased, Mars lost not only its opportunity to have liquid water but also its ability to remain relatively warm at night. This increased the rate at which water disappeared from the Martian surface, driving it into either the permafrost or the polar caps.

The net effect was to turn Mars from a primitive planet with liquid water to a totally dry world, devoid of any chance of liquid. This change probably took no longer than a few hundred million years, at a time somewhere between three and four billion years ago. During that era, it was good-bye to a Mars fit for life, and hello to a planet hostile to life as we know it.

What we know and have concluded about the origin of life on Earth strongly suggests that primitive Mars may well have been nearly as ready for life as primitive Earth was. If life did begin on primitive Mars at a time close to four billion years ago, the burning question that confronts us today is, Did life totally disappear on the Red Planet—or did life find a few places of refuge, where even today it might persist, awaiting our discovery?

If life requires liquid water, as all life on Earth does, then we cannot expect to find life on the Martian surface. But this does not eliminate all the possibilities for Martian life. We can envision forms of life that have found niches beneath the polar caps, or, even better, in subterranean caverns heated by volcanic activity. These "underground oases" are entirely imaginary—for now. But

Can Life Exist on Mars Now?

The south polar cap of Mars consists of frozen carbon dioxide plus some ice. Though no water, necessary for life, can exist on the surface of Mars, scientists speculate that organisms may survive under the frozen polar caps. (*NASA photograph*)

they rank high on the list of items that scientists would like to investigate in any return to Mars. Temperature-sensing satellites might be able to locate underground regions significantly warmer than their surroundings, and robot explorers might then be able to drill into what might prove underground reservoirs where forms of life have survived nearly four billion years despite the absence of liquid water on the Martian surface.

Enough of these "mights"! Think of the excitement and the frustration of scientists who have dreamed of finding life on Mars, have lived through the twenty years that followed the negative results from the *Viking* missions to Mars, and have now learned from ALH 84001 that life may have existed on Mars three to four billion years ago. Only a few dozen million miles separate them from the chance to determine whether any life exists on Mars now, and if it does, to analyze those forms of life to discover their differences from, and resemblances to, Earth life.

How Different Should We Expect Martian Life to Be?

The analysis of ALH 84001 could not hope to reveal any DNA, or anything like it, since no such molecules can last for billions of years without degrading—that is, falling apart into much smaller molecules. The degradation of DNA starts immediately upon exposure to the outside environment, which is why police detectives are supposed to lose no time in taking their samples to the laboratory for analysis. But what about the fictional dinosaur DNA made famous in the book and movie *Jurassic Park*? This scenario imagined that a mosquito that had bitten a dinosaur became trapped in amber, which preserved enough dinosaur DNA for scientists to reconstruct entire animals. Scientists believe that DNA might have a chance of actually surviving in this manner, but that it is highly unlikely, no matter how many mosquitoes drowned in amber we may eventually find and examine.

Eventually, however, we may find living organisms

on Mars, or on Jupiter's moon Europa, or elsewhere in the universe. If these are microorganisms, as seems highly probable from the example that Earth provides, then apart from the enormous implications of their existence, the single most meaningful question that these organisms can answer will be, How similar to, and different from, Earth life's chemical reactions are those that other forms of life employ? Do they, for example, all have the same means of replication? Does that means also rely on a DNA-like molecule that governs the formation and operation of new cells?

Suppose, for instance, that we find extraterrestrial life and discover that it uses exactly the same sort of DNA that Earth life does. Nothing could point more strongly toward a common origin for the two forms, and thus toward the cosmic seeding that we discussed in the previous chapter. On the other hand (and most biologists would vote for this as the more likely possibility), all extraterrestrial forms of life—if they exist!—may employ notably different chemistry, along with significantly different types of molecules to govern the operation of their "cells" and the replication that creates new organisms. These differences would bring into being a new science of comparative biology, not among various types of Earth life but between and among truly different forms of life. "I hope that if we do find life, it *is* different," says Jack Szostak, a molecular biologist at Massachusetts General Hospital who studies the origin of life. "If not, that would be truly disappointing."

We might find, for example, that the first extraterrestrial forms of life to be discovered replicate by using a molecule that is similar in its structure to DNA, but forms its spiral backbones and cross-linking pairs from molecules that differ from Earth life's DNA. This would strongly support the notion that most forms of life employ a similar set of processes and a similar molecular structure; the differences in the details would reveal a great deal about how life's evolution seized on what was available to achieve its ends.

The reader may well suspect that I have fallen into a common trap, one that we must always attempt to shun: imagining that other forms of life strongly resemble our own. Similar DNA, indeed! What about extraterrestrial life that has nothing like DNA, nothing like the fluid-filled cells of Earth life? These objections have complete validity. We must in fact guard against the assumption that extraterrestrial life resembles our own, and ourselves in particular. This false assumption, the bane and nemesis of UFO enthusiasts, must be fought at every turn. But extraterrestrial life *might* turn out to employ a chemistry similar to Earth life's, and if such life consists of microorganisms, we are likely to recognize that type of life more readily than the forms that differ far more widely from Earth life. We can collapse all objections readily by admitting that the most important step in resolving the question of how much, and in what ways, extraterrestrial life resembles Earth life will occur when we find other forms of life. The first example of extraterrestrial life will tell us a great deal about the validity of the panspermia hypothesis, and half a dozen examples will take us well down the road that leads to an understanding of the basic principles that govern life's origin and development on planet after planet.

1976: The Search for Life on Mars

The preceding paragraphs sketch the theoretical arguments against life on the surface of Mars. These arguments carry more weight now than they did before we sent spacecraft to Mars, because the thinness of Mars's carbon-dioxide atmosphere had been imperfectly established. Besides, as every scientist knows well, theory plays a crucial role in advancing scientific knowledge, but not the final one, which rests with experiment. So astronomers and biologists designed two marvelous spacecraft to visit Mars, orbit the planet, send landers to its surface, and look for signs of life in the Martian soil.

In designing these experiments, the scientists looked at our planet's most Mars-like environment, the coldest and driest regions on Earth: the Dry Valleys of Antarctica. There, a few dozen miles from the main United States research base, the mountain patterns and prevailing wind directions keep the precipitation below half an inch per year, almost all of it in the form of light snow. Unlike the rest of Antarctica, which lies beneath a two-mile-thick sheet of ice, the Dry Valleys, also called the Ross Desert, are essentially free of ice. And there, on the inhospitable slopes above the valley floors, the biologist Imre Friedmann discovered tiny organisms named cryptoendoliths ("hidden inside the rocks").

Cryptoendoliths are actually complex colonies of symbiotically interacting lichens that find shelter just below rock surfaces from the harsh temperatures, and sudden temperature changes, that rule the outside world; they receive barely enough sunlight, filtered through the porous rocks, to survive, and not enough to thrive. They represent "life at the edge," organisms that remain alive only by immensely retarding the rates of their metabolic processes: An ordinary-looking rock may contain a cryptoendolithic colony thousands of years old. In describing life (not to mention research) under these hard-pressed circumstances, Imre Friedmann quotes Dante's *Inferno:* "So bitter it is that death is hardly more so; but to treat of the good I found in it, I shall tell of the other things I saw." Even though the Dry Valleys furnish us with Earth's most Mars-like locale, the real Mars must be harsher still, with its thin carbon-dioxide atmosphere and its zero—not merely microscopic—annual amounts of rain or snow.

The Viking *Life-Detection Experiments*

When astronomers and biologists began to plan the *Viking* experiments to search for life on Mars, they knew that they faced an enormously difficult task in designing a miniature robotic laboratory that could survive a journey of a hundred million miles, land and deploy itself on a strange planet, and then send the results of its experiments back to Earth. Prolonged discussions led to the design and construction of the two finest miniature laboratories ever built, which reached Mars in the summer of 1976, separated from the *Viking* orbiters, and landed safely on the planet's surface, avoiding impalement on any of the numerous rocks that would have destroyed them. For two years, these *Viking* landers radioed thousands of pictures from the Martian surface, record-

Can Life Exist on Mars Now?

ing the cycle of days and nights, summers and winters, which left carbon-dioxide frost on the landscape as the temperatures fell to 100 degrees below zero.

None of these images showed a single trace of life. Nothing like a living creature or anything resembling a trail, a burrow, a corpse, or the waste products from a terrestrial organism, appeared in the images. By itself, this meant relatively little, for, as we have seen, most life on Earth is microscopic in size, utterly invisible to the

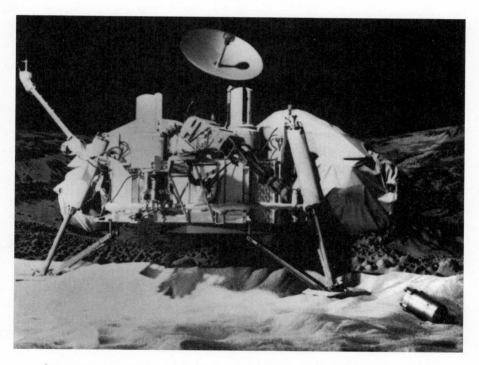

An engineering model of one of the *Viking* landers appears against a painted backdrop. For two years the *Viking* landers sent thousands of pictures from the Martian surface, providing a remarkable close-up view of the planet. (*NASA photograph*)

The Hunt for Life on Mars

This view of Mars's surface taken by the *Viking 1* lander in 1976 shows a barren landscape of rocks and dust, but without a single indication of life. (*NASA photograph*)

unaided eye unless present in enormous numbers of individuals. This was the reason for spending hundreds of millions of dollars on the robotic laboratories, which could test for the presence of life too small to appear before the *Viking* cameras.

The miniaturized laboratories included no microscopes, which had been judged too expensive for the relatively small potential benefit, in comparison with the three experiments that did make the cut. In other words, the *Viking* scientists placed their money not on a visual inspection of the Martian soil but rather on a search for *chemical* evidence for the existence of life in that soil. What type of chemical evidence would prove most convincing? This question plagued the researchers, and has hung over the discussion of the *Viking* results for two decades. The controversy highlights the fact that deciding what constitutes proof of life—indeed, what constitutes life itself—cannot be settled to everyone's

Can Life Exist on Mars Now?

satisfaction. The further we get from familiar forms of life, the more difficult the definition. Several years of debate made the *Viking* scientists entirely familiar with this problem, so that their final decisions in constructing the laboratories, and their long wrangles over how to interpret the results of their experiments, attest not to any lack of awareness but instead to the enduring nature of the fundamental question, What is life?

What were the three *Viking* chemical experiments? Broadly speaking, one looked for the results of any *breathing* on Mars, the second aimed to find microorganisms in the process of *eating*, and the third sought to roast the *corpses* of any Martian microbes and to detect their existence by the gases they released. The scientific names for these three searches were the labeled–release (breathing), gas–exchange (eating), and pyrolitic–release (corpse–roasting) experiments. The breathing experiment dripped liquid onto a sample of Martian soil, carrying a set of chemical compounds that had been "labeled" with radioactive carbon atoms, which could easily be recognized by a detector in the miniaturized laboratory. If gases containing the radioactive carbon, such as carbon dioxide, then appeared above the soil, this would be a sign that microbes in the soil had "breathed out" some of the labeled compounds. In a similar way, the eating experiment dropped Martian soil into a broth of several dozen "likely nutrients" and searched for changes in the gas immediately above the broth. Any changes would show that microbes had incorporated at least some of the liquid nutrient and had produced new gases as a result.

But what if Martian microbes, unexposed to water for billions of years, had no use for the liquid the experiment dripped onto the soil, and therefore "breathed" nothing out? And what if the broth of liquid nutrients—called "chicken soup" by the scientists—had no appeal for Martian organisms? The corpse-roasting experiment dealt with these difficulties by placing Martian soil into a chamber where the atmosphere duplicated that on Mars, except that the carbon atoms in the atmospheric carbon-monoxide and carbon-dioxide molecules were "tagged" with radioactive isotopes. After a few hours of exposure to this atmosphere, the soil was heated to more than 1,000 degrees Fahrenheit (well above 500 C), which pushed the mock Martian atmosphere, along with the gas released by any roasted corpses of microorganisms, into a "vapor trap" that captured these gases (except for the "tagged" carbon monoxide and carbon dioxide, which passed completely through the trap). The gases caught in the vapor trap were then measured for any radioactivity, which could only have come from radioactive carbon compounds incorporated by the organisms while they were alive.

With these three experiments, *Viking* had a good chance of finding any organisms that breathed, ate, or left corpses with organic material when they died.

Did the Vikings *Find Life on Mars?*

In 1976, a few weeks after the *Viking* spacecraft had landed on Mars at two sites separated by several thousand miles, the results reached Earth: In both locations, all three life-detecting experiments showed positive results! Yet few of the scientists concluded that microbes exist on Mars. Instead, they recognized, more suddenly than foresight would easily allow, the immense difficulties involved in distinguishing chemical changes caused by life from those caused by nonliving processes—especially when this distinction aims at finding an unknown form of life.

The eating experiment, first to report its results, showed that significant amounts of oxygen appeared above the Martian soil soon after the nutrient-rich "chicken soup" dripped onto it. But a consensus soon emerged that instead of producing this oxygen by biological processes, the Martian dirt had responded to the presence of liquid with chemical reactions that released some of the oxygen locked in the soil. In other words, the increase in humidity caused by introducing the liquid nutrient by itself led to chemical reactions that released oxygen. The scientists told the robot laboratory to repeat the eating experiment, this time using Martian soil that had been preheated to a temperature above the boiling point of water, considered sufficient to destroy any Martian form of life. When oxygen appeared once again, the conclusion seemed firm: Dropping liquid onto Martian soil could release oxygen without the action of any biological processes.

The breathing experiment reported a day or so later, and again the results were positive. After a few drops of the labeled compounds were dripped onto the Martian dirt, the counter showed an immediate increase in the number of radioactive carbon atoms in the gas *above* the soil—an increase more dramatic than had been found during test experiments with many types of life-bearing soil on Earth! But the scientists realized that many compounds quite likely present on Mars, such as hydrogen peroxide, would release carbon dioxide once the nutrient was dripped on them. They told the laboratory to try a second drip, and it reported no increase in the radioactivity. If microorganisms were present, they should have "breathed" as deeply in the second experiment as in the first. The scientists therefore concluded that chemical reactions without biology could have produced, and almost certainly had produced, the positive signal from the experiment.

And what of the relatively unfoolable corpse-roasting experiment? This one involved no liquids, but was simply designed to see whether radioactive carbon atoms in the simulated Martian atmosphere became part of the Martian soil put into the chamber. Indeed some did: Although the signal was weak, the corpse-roasting test showed that *some* carbon atoms had entered the soil.

Can Life Exist on Mars Now?

Once again the *Viking* scientists told their faraway laboratory to repeat the experiment, this time with Martian soil that had been preheated for three hours, in one case to 200 degrees F (93 C) and in a second case to 350 degrees F (180 C). The lower amount of preheating had no effect; the second reduced the number of carbon atoms taken into the soil by 90 percent. No one had envisioned Martian organisms that could withstand temperatures of 200, let alone 350 degrees of heating. Hence the biologists concluded that the incorporation of some carbon atoms into the Martian soil had occurred by nonbiological chemical processes.

The three *Viking* experiments designed to register signs of life thus all proved to be chemistry experiments. If chemical reactions can do all this on Mars in the absence of life, how can we ever test another planet for the existence of microbial life with certainty? The *Viking* experiments show the difficulty of this effort, and the discoveries of 1996 reinforce it, since even the finest earthbound laboratories cannot yet answer the question of whether or not ancient life existed in the meteorite ALH 84001.

Can any purely chemical evidence prove the existence of life? Bill Schopf says no. At the press conference on August 7, 1996, Schopf specified the "smoking gun" that he would want to see before pronouncing that life exists. He wants to see cell walls; he wants to see data that show an entire population of organisms, with a range of sizes and shapes; and he wants evidence of cell division. And how much of this evidence does he require? "Give me five hundred [cells that show cell walls]; that's enough," he said in August 1996. The edges of the carbonate globules in ALH 84001 may eventually yield this sort of evidence, but the *Viking* landers, which lacked any sort of microscopes, obviously could not provide it, whether or not that evidence exists on Mars.

Both the search for life on the Martian surface and the analysis of ALH 84001 cry out for additional evidence, either in support of or in opposition to the hypothesis of living organisms. The *Viking* experiments did include another, possibly crucial instrument that measured the amounts of different types of molecules in the Martian soil and atmosphere. This instrument, called a gas chromatograph–mass spectrometer, showed that Martian soil contains less than one part per billion of any compounds that would be called "organic"—carbon-based molecules, like the molecules found in living organisms on Earth. Even Antarctic soil, far from any living creatures, contains a small but still easily detectable amount of organic material, which arises either from the decay of long-vanished creatures or has floated down from the atmosphere. The failure to find any organic compounds in the Martian soil weighs in the balance against the existence of life on Mars, though, by itself, does not disprove it.

What do the results of the *Viking* landings on Mars tell us? Almost all the biologists who have looked carefully at the *Viking* results conclude that chemistry on Mars duplicated the biological activity that the miniature laboratories had been designed to detect, and that the Martian soil sampled by the landers contains no life at all. These experts might, of course, be wrong. One well-known member of the *Viking* team, the biologist Gilbert Levin, has concluded that Martian microbes do offer the most likely explanation for the results from the *Vikings'* life-detection experiments; after all, they each returned a positive signal. But all of Levin's cohorts maintain that the opposite is true: that the positive results from each of the three experiments find their best explanation in chemical reactions that mimic the activity of microorganisms on Earth.

The fact that the *Viking* experiments gave positive results, yet were interpreted to show the absence of life, has dogged NASA for two full decades, prompting ever-popular assertions, still present in the tabloids, that scientists actually discovered life on Mars in 1976 but that NASA has sought ever since to conceal their findings. Just *why* NASA should do so rarely receives an airing. However, those who find it a reasonable conclusion that the United States government would keep alien corpses frozen in Ohio may well find nothing odd in the notion that NASA would attempt to consign to oblivion the positive results from its most ambitious effort ever in the search for extraterrestrial life—results that could have assured increased funding well into the new millennium.

Looking to the Future—and to the Past

After 1976, after *Viking 1* and *Viking 2* had engaged in an exciting but ultimately fruitless search for life on the surface of Mars, a pall fell over the quest for extraterrestrial life. Even though scientists knew that the *Vikings* had sampled only a bit of the surface layer close to the two landing sites, that the absence of liquid water on

Mars made life there a long shot, and that they had to develop plans to reach the polar caps or to drill underground, the ineluctable fact remained that Mars, the most promising candidate in the solar system for life beyond Earth, had come up empty. For the past twenty years, Mars has seemed a dead planet.

All that changed with the announcements of 1996. R.I.P. dead Mars, viva Mars with life—even if that life existed billions of years ago! Although ALH 84001 has not provided clear proof of life and may never do so, the ancient Martian meteorite confirms the evidence for ancient river beds by showing that Mars apparently *did* have running water in its far-distant past, and for the first time we have in hand some of the material—the carbonate globules—that formed in the presence of this water. The scientists who dreamed of returning to Mars with better knowledge and improved techniques have revised their projects as they have assimilated the new information. We can now take a mental excursion with them to the cold, dry plains of Mars.

"Can we?

Should we?

Will we?"

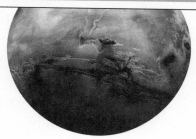

FUTURE
MISSIONS
TO MARS

The news about ALH 84001 fell like water on a parched soil for those who dreamed of returning to Mars to search for life on the Red Planet. As luck would have it, at the time that the news broke, NASA had two automated space-craft almost ready for launch to Mars and Russia had a third. Although these spacecraft will not answer all the basic questions about the possibility of ancient life on the planet, they should increase our knowledge and end a long hiatus in our exploration of Mars.

Waiting for the Proper Alignment

All three spacecraft had to be launched during the last two months of 1996 in order to take advantage of a favorable lineup of Mars and Earth, one that provides the least expensive trajectory (in energy terms, and thus also in real dollars) from our planet to Mars. These lineups recur at intervals of 780 days, when the two planets regain their same orientation with respect to the sun. This happens because in 780 days the Earth "laps" Mars once, making 2.14 complete trips around the sun during the time that Mars, moving in a considerably larger orbit at a speed only 81 percent of the Earth's, covers just 1.14 of its orbits. This means that if an alignment favorable for launch occurs today, a similar opportunity will present itself two years and seven weeks later. The next "launch window" for missions to Mars will occur as 1998 turns into 1999, and NASA even now is redoubling its efforts to design and build spacecraft that can best take advantage of this opportunity.

The three spacecraft launched in 1996 had been designed with the key result from the *Viking* landers firmly in mind: Life does not seem to exist on the Martian surface now. As a result, all three missions have only limited abilities to search for life, but they have been well designed to provide more information about Mars and its evolutionary history. The missions will also provide a useful benchmark for assessing the difficulties of sending a spacecraft to Mars that would have a reasonably good chance of finding life there—if life exists. Let us look at what these spacecraft can achieve before 1997 ends, along with the much greater tasks that they will leave for later explorers to accomplish.

The Two NASA Missions:
Mars Pathfinder *and* Global Surveyor

NASA's two missions to Mars have difficult functions aimed at achieving separate goals. *Mars Global Surveyor*, launched at the end of 1996, will reach Mars in September 1997. Using the thin Martian atmosphere to slow itself down, and firing rockets to adjust its trajectory when it reaches the planet, *Global Surveyor* will enter an orbit around Mars at only a few hundred miles above the Martian surface. There the spacecraft will be well placed to map the surface, measure the temperature and composition of the atmosphere, and send these data back to Earth by radio.

NASA hopes that *Mars Global Surveyor* will compensate in large part for one of the great disasters in planetary-exploration history, the loss of NASA's *Mars Observer*, a carefully designed spacecraft that reached Mars in perfect condition in August 1993 and then lost radio contact with Earth. Despite all attempts to reestablish a radio link, the spacecraft was never heard from again. The loss of *Mars Observer* came four years after the Soviet Union lost two spacecraft, named *Phobos I* and *Phobos II* after one of Mars's two small moons, which they were meant to explore with laser beams and landers. Ground controllers in the Soviet Union lost contact with *Phobos I* while it was still on its way to Mars, apparently as the result of a computer-programming error, and *Phobos II* stopped sending signals to Earth almost immediately after it had achieved an orbit around Mars.

"For individuals not directly involved in space missions, it's hard to appreciate the anguish felt [when missions fail] by the men and women who devote major portions of their careers to these enterprises," says the University of Hawaii's Tobias Owen, who has himself spent large amounts of time as a participant in such missions. "Everything those people do is charged with anticipation: years of design and development, new inventions, subtle software, carefully tested experiments. These are the most sophisticated devices our civilization can produce that are not designed to kill someone. Their purpose is to reveal the secrets of other worlds. When one of these marvelous messengers from Earth falls silent, the letdown can be overwhelming. After the tears and depression, a certain gallows humor began to emerge, and there was talk of a 'great galactic ghoul' that lived near Mars and liked to devour spacecraft, or had been annoyed by the *Viking* orbiters and landers in 1976." It is easy to believe that when scientists and engineers themselves indulge in such talk, the popular press is not slow to assert the reality of the professionals' demons, which in fact are engineering and technological failures within highly sophisticated machinery. "After we lost contact with *Mars Observer*," Owen recalls, "popular paranoia produced a strange demonstration outside JPL [the Jet Propulsion Laboratory in Pasadena, California, which manages the spacecraft that NASA sends to other planets]. A group of people had become convinced that the mission had actually succeeded and in fact had found evidence of advanced civilizations on Mars that NASA was concealing from the public.''

Ever since Owen had noted a feature resembling a human face on one of the *Viking* orbiters' photographs, the assertion that an ancient race of highly developed Martians had left behind a huge monument had become deeply rooted among certain sectors of the public. Actually, the "Face on Mars," like the "Man in the Moon" or the Great Stone Face in New Hampshire, testifies to our brains' ability to "recognize" familiar shapes, especially faces, in random assortments of light and shadow. With any luck, *Mars Observer* will obtain high-resolution im-

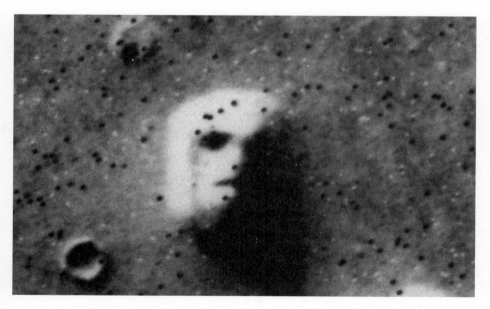

One of the photographs of the Martian surface taken by the *Viking 1* orbiter in 1976 shows a rock formation about a mile across that reminds some observers of a human face. (*NASA photograph*)

ages of the "Face on Mars," clearly demonstrating that the Martian topography has fooled those who wish to be fooled. This evidence could change the minds of those who believe that NASA is engaged in a massive cover-up, though they may well demand further proof.

In any case, the three spacecraft lost on the way to Mars or in orbit around the planet dealt a multiyear setback to increasing our knowledge of the planet. "The net result of the three spacecraft failures was certainly depressing," Owen recalls. "By the time ALH 84001 was analyzed, twenty years had gone by since the *Viking* mission, which had been so highly productive, and in all that time no spacecraft had successfully reached Mars. For comparison, going back in time twenty years *before* the *Viking* landings takes us to 1956, one year before the Soviet Union launched the first artificial satellite of Earth. The pace of exploring the solar system slowed dramatically during the 1980s and early 1990s, despite the launch of the *Galileo* mission to Jupiter in 1989. Fortunately for those of us interested in Mars, this dry period is about to end."

On July 4, 1997, two months before *Global Surveyor* enters orbit around Mars, *Mars Pathfinder* is scheduled for landing on the Martian surface. Launched in December 1996, *Mars Pathfinder* will deploy a parachute to allow the Martian air to slow its descent. Still falling at a thousand miles per hour, the spacecraft will inflate four 15-foot-wide air bags just before encountering the surface. Together with retro-rockets that fire briefly to slow the craft still further, the air bags should allow *Mars Pathfinder* to bounce thirty feet high and achieve a soft landing on one of Mars's ancient flood plains. This region, Ares Vallis, satisfies the dual requirements of offering a relatively flat landing surface and placing the lander close enough to the Martian equator to receive sufficient sunlight to run its solar-powered cells. Eventually, the NASA experts who pick sites to search for possible fossils would like to visit the Dao Vallis region, where they have identified what seems to be a valley in which warm water gushed from hot springs during the long-vanished era when Mars had running water. But Ares Vallis (see photograph on page 173) remains an attractive alternative; if all goes well, *Mars Pathfinder* will land somewhere within a 60-by-120-mile ellipse in this region.

After *Pathfinder* deflates its air bags, its small motors will set the spacecraft upright and open the protective covers of its scientific instruments and communications antennas. During the Martian day, *Pathfinder* will use its cameras to survey the landscape and to distinguish the different types of rocks that it sees; it will also employ meteorological instruments to check on the weather. Most significantly, *Pathfinder* will unfurl twin ramps, down one of which an electromobile will roll—the first self-propelled vehicle sent from Earth to survey the Martian surface.

Great projects begin with small advances. The "rover" that *Mars Pathfinder* carries measures a foot and a half long by two feet high, not much larger than the model cars once raced by intense adolescents. This vehicle, with a solar-powered electric motor and a six-wheel pantagraphic suspension system to master the relatively rugged terrain, costs a good deal more than any of Detroit's finest. Accepting a suggestion made by a twelve-

On the flank of an ancient Martian volcano lies a large channel called Dao Vallis, visible at the lower left of this *Viking* orbiter photograph, into which water apparently flowed and where hydrothermal springs may have existed, many billion years ago, leaving behind deposits that geologists call sinters. (*NASA photograph*)

An artist's rendering of the *Mars Pathfinder*'s descent to the Martian surface shows a cocoon of airbags that will soften the landing. Once the spacecraft stops bouncing, the airbags will deflate and the spacecraft will stand itself right side up, then open its "petals" to expose its solar panels.

year-old schoolgirl, NASA has named the rover *Sojourner* in honor of Sojourner Truth, the famous black evangelist of the pre–Civil War era who was born a slave in New York but was freed by her owner just before the state abolished slavery in 1827. After waging a successful court battle for her son's freedom (he had been sold illegally to Southern slaveholders), she changed her name and preached the abolition of slavery and the right to vote, especially to meetings of woman suffragists.

Every day on Mars will bring an opportunity for *Pathfinder* to identify interesting sites for *Sojourner* to examine. Then, guided by commands from Earth, *Sojourner* will roll up and down over the Martian terrain, reaching distances as great as five to ten yards from the lander. If *Sojourner* meets a rock too large to climb over, it can maneuver around it, but it must maintain line-of–sight contact with the lander to function properly.

Sojourner's mission is to roll up to local rocks that have been deemed promising from pictures transmitted by the lander and attempt to determine their composition. To do so, the rover carries efficient miniature cameras to photograph the rocks, as well as a device to chip away at them and thus to allow earthbound geologists a chance to scrutinize newly exposed rock surfaces on Mars. In addition, *Sojourner* can shoot a spring-loaded spike into rocks to test their hardness, and carries a spectrometer that will yield additional information about the rocks' composition.

Like *Global Surveyor*, *Mars Pathfinder* represents the new era at NASA, governed by agency administrator Daniel Goldin's motto "cheaper, faster, better." (Cynics sometimes compare this to a sign much favored by auto

The open petals of the *Mars Pathfinder* are covered with solar cells, which will provide electrical power. The gold thermal blanket will protect the computer and other electronic equipment from the temperature extremes on Mars. The miniature rover, *Sojourner*, will roll down the exit ramp onto the surface of Mars and conduct a week of scientific experiments.

mechanics that invites the customer to choose any two of those three criteria.) The cost of these spacecraft, adjusted for inflation, falls far below that of the *Vikings*, which had to achieve a soft landing with retro-rockets, a far more expensive task than deploying a parachute and air bags, and carried the most marvelous miniature laboratories ever manufactured on Earth. NASA hopes that *Mars Pathfinder* will show that even though Mars's atmosphere is less than one percent as thick as Earth's, soft landings without rockets are nevertheless feasible. Similarly, *Global Surveyor* should demonstrate that we can send automated instrument platforms into orbit around Mars at low cost.

The Sad Fate of the Russian Entry: Mars 96

Russia's automated spacecraft, *Mars 96*, was scheduled to reach Mars by the end of 1997, achieve an orbit around Mars, and then send two landers down to the Martian surface. In addition, the orbiter was to launch two "penetrators," probes to be fired into the surface at high velocity, survive the impact, and penetrate the Martian soil to a depth of several feet. These penetrators, had they functioned properly, would have made the first measurements of the Martian surface at depths below a few inches. They would have also placed cameras and meteorological instruments on the surface directly above, to transmit additional data concerning these possible landing sites on Mars.

On November 16, 1996, when *Mars 96* was launched from the Baikonur Cosmodrome in Kazakhstan, the final stage of its rocket failed to ignite. As a result, the spacecraft plunged ignominiously into the Pacific Ocean, drowning its $200 million payload, the product of many years' labor by scientists and engineers in twenty countries. Because the Russian space program uses rockets more powerful than any at NASA's disposal, the instruments of *Mars 96* weighed considerably more than those on *Mars Pathfinder* and *Global Surveyor*, and could have provided greater sensitivity, especially in the camera system. Those instruments now sleep in the briny ocean, one more entry on the list of spacecraft that have failed either to reach Mars or to perform properly once they arrived. The *Mars 96* disaster may put an end to the Russian space effort, which had already been poorly funded and will now have a difficult time finding customers for its launch vehicles. Scientists must, however, attempt to overcome this negative result and continue to plan their future successes.

The Next Generation of Automated Spacecraft to Mars

All three of the 1996 spacecraft were designed and constructed before the news from ALH 84001 broke. Even the *next* set of missions to Mars, those scheduled to be launched at the end of 1998, have already passed the point where any significant alterations are possible. Fortunately, the most important mission, called *Mars Surveyor*, was already well designed to increase our understanding of Mars and its possible habitability.

NASA plans to launch *Mars Surveyor*, a mightier instrument than either *Global Surveyor* or *Mars Pathfinder*, in December 1998. Like the *Vikings*, *Mars Surveyor* will include both an orbiter, to study the planet from above and to relay data back to Earth, and a lander, which can make detailed investigations of a small area on the surface. Unlike the *Vikings*, however, *Mars Surveyor*'s orbiter and lander will make separate journeys to Mars. This plan allows NASA to use relatively inexpensive Delta II rockets for each trip. Also unlike the *Viking*, *Mars Surveyor*'s lander will aim not for the broad plains of Mars but for a spot near the edge of the south polar cap, where the lander will cut carefully into the soil, exposing successive layers whose images will be sent to Earth. During the same launch window, at the end of 1998 or the beginning of 1999, the Japanese space agency will launch the *Planet-B* spacecraft, designed to go into orbit around Mars and obtain more information about its atmosphere. To achieve this goal, *Planet-B* will carry a spectrometer built by United States scientists to make detailed measurements of the atmospheric composition.

Enough of the Pictures and Measurements: What About Bringing Samples Back to Earth?

The debate over whether or not the evidence in ALH 84001 proves that ancient life existed on Mars poses this challenge: If we hope to find definitive proof of life, we probably need to obtain not just photographs or measurements of the rocks on Mars but actual *samples* of them, to be studied in detail in terrestrial laboratories. The analysis of ALH 84001 illustrates the impressive fineness of detail that sophisticated laboratories on Earth allow scientists to achieve, using equipment capable of counting tiny numbers of radioactive nuclei or performing intricate chemical analysis with the application of lasers and electric fields.

Despite the impressive achievements of automated laboratories such as those on the *Viking* landers, we may yet find that even the latest generation of these miniaturized marvels cannot find life on Mars, or traces of fossil life, if we need to resolve details similar to those seen by David McKay, Richard Zare, and their collaborators. We have been fortunate to have had the chance to collect and identify a dozen rocks from Mars, at an expense less than a thousandth of sending even the simplest spacecraft to another planet. But further research on the Martian meteorites may not significantly improve our understanding of the possibilities of ancient life on Mars, and even roving spacecraft on the Martian surface may not yield definitive proof or disproof of the present existence of life on the Red Planet.

We may note with regret that the instruments we send to other planets with our finest spacecraft are almost by definition obsolete by the time they arrive. Not only does the journey take the best part of a year (for trips to Mars or Venus) or even several years (for travel to Jupiter, Saturn, and the other giant planets); more important, the spacecraft must be designed, engineered, constructed, and then thoroughly tested before being delivered for launch. If a launch window closes before the spacecraft makes a successful takeoff, more than two years, at least in the case of Mars, may pass before the window reopens.

On the other hand, if we can bring samples back to

Earth, the samples *never* become obsolete. In fact, as scientists develop new techniques for studying them, the samples effectively become *more* up-to-date in the information they can yield. ALH 84001 provides a perfect example of this fact, even though it came to Earth by accident rather than human endeavor. As scientists improve their techniques for scrutinizing a meteorite such as this one, and recognize that additional tests can be performed on samples they have already studied, they can draw new information from old rocks. However, ALH 84001 is a "tough rock," made by volcanic activity and congealed into something like the igneous rocks of Earth. What astrobiologists really want to study are *sedimentary* rocks, those formed over time by deposits underwater. On Earth, these are the fossil-bearing rocks; even expert paleontologists cannot find fossils in igneous rocks. But sedimentary rocks are too fragile to reach Earth by the impact method. Any object that blasts free these rocks—if they exist on the surface of Mars—will destroy them, shattering them into small bits that will not be recognizable as meteorites if they should happen to fall on Earth.

What may prove necessary, and what certainly excites those who plan future missions to Mars, is the possibility of combining automated rovers that will locate likely sites for Martian life with return-trip vehicles that will bring samples of the Martian surface and subsurface back to Earth. According to NASA's current plans, this potent combination cannot be ready for the 1998/1999 or 2001 launch windows; but the next opening, for launch to Mars in mid-2003, arrival in early 2004, and a sample return by 2005, may be achievable if NASA receives additional funding for this venture. In any case, the first decade of the new millennium should see rocks brought back from Mars, provided that enthusiasm for exploring the Martian surface remains strong among those who would fund such an endeavor. When this occurs, no one will be more quietly satisfied than the man who has devoted much of his life to the quest for sample returns, Christopher McKay.

The Quest to Return Samples from Mars

Chris McKay, the second key McKay we encounter in this story, is unrelated to David McKay, who led the team

Chris McKay, an expert in astro-geology, dreams of working with samples of Mars brought back to Earth for detailed examination. (*Photograph by Donald Goldsmith*)

that investigated ALH 84001. A tall, rangy man in his forties, McKay received a Ph.D. in "astrogeology" from the University of Colorado fifteen years ago, and joined the NASA/Ames Research Center in California to study the relationship between the evolution of the solar system and the origin of life. McKay has traveled to both polar regions in search of Mars-like conditions: He has climbed up and down the sides of the Antarctic Dry Valleys to aid Imre Friedmann in his search for the cryptoendolithic life that hides in the rocks, and plunged in low-temperature scuba gear into Lake Hoare, a large pond on the floors of one of the Dry Valleys, to investigate the algae that manage to live beneath the lake's permanent ice cover.

"An ice cover is a good thing to have, since it keeps the rest of the lake from freezing," McKay notes. "If ice were heavier than water, none of that could happen"— another reason why water makes an excellent solvent for any form of life. "You know, Europa [one of Jupiter's four large satellites] has an icy surface. The *Galileo* photographs [taken in 1996 by the spacecraft that went into orbit around Jupiter earlier that year] seem to show cracks in the ice, that could connect with water below. That would be a great world to investigate to see whether prelife conditions, or even life, could exist there."

But while Europa has its appeal, what makes McKay's eyes light up most is the thought of going to Mars, initially with automated rovers, then with spacecraft that can bring samples back to Earth. "ALH 84001 tells us two key things about a sample-return mission," says McKay. "First, the fact that the *first* ancient rock that we found from Mars had been exposed to liquid water means that just about any rock on Mars's cratered terrain should be interesting. It confirms our notion that early Mars was immensely different from Mars today, which is dull as beans. We don't have to spend two decades planning where to go for a sample return if the very first ancient rock is full of great stuff. ALH 84001 shows that at times between 4.5 and 3.5 billion years ago, Mars was wet, warm, and reducing [endowed with an atmosphere rich in hydrogen]. The processes all over ancient Mars apparently put a crust, a layer of varnish, all over the rocks on the surface, locking in the evidence. Every crack and crevice in the ancient rocks on Mars should preserve their evidence of life, if it exists.

"Second, ALH 84001 shows that the cratered terrain on Mars is an important site to search for life. There are large regions in this terrain that have not been disturbed for more than three and a half billion years." Astronomers deduce this fact from the terrain's large numbers of craters, which are presumed to have arisen during the later stages of the era of intense bombardment after the solar system formed. "It's not trivial to land in the cratered highlands, both because the surface is so rugged and because it's more than a mile higher than the low lands, so there's even less atmosphere to brake the spacecraft, but that's the best place to get a sample. The next best places are the ancient lake beds and the streams that fed them, and they are easier places to land in. Actually, the lake beds are great places to search for fossil life on Mars. There are two practical reasons: They're big and flat, so we can identify them with certainty. And a lake bed is good for fossils—they're basically independent of the model according to which life would leave fossils behind. Finally, the Antarctic environment suggests that life can persist in a lake under an ice cover as the climate dries out."

Though the news from Mars in 1996 has stimulated NASA to new efforts in planning a sample-return mission, it would be premature to name a date when that will occur. "The first rock was free, but the next ones will cost about $300 million," says McKay. "We should plan for a sample return soon. It can be quick and fast and easy—even if we just grab a quick handful of rock and dust." This is the sort of mission that would cost $300 million; more advanced spacecraft would cost significantly more. "Sample returns are expensive," McKay points out, as he well knows from many discussions.

What About Humans on Mars?

When most people think about investigating Mars, their image revolves around a *human* presence on the planet. Automated landers, rovers, and explorers capable of photographing the surface, drilling into the subsoil, or examining the edges of the polar caps may all be quite useful, but surely their chief project must be to prepare the way for human explorers. Has this not been the history of human interaction with expanding horizons? And must our destiny not lie in exploring and colonizing our neighboring planets?

When Ray Bradbury, author of *The Martian Chronicles*, wrote about the news from Mars, he well expressed this fundamental human belief that only personal visits seem worth the effort. "The apparent discovery of life on the Red Planet," said Bradbury in 1996, "is only worth our hyperventilation if we allow it to lead us to the larger metaphor: mankind sliding across the blind retina of the Cosmos, hoping to be seen, hoping to be counted, hoping to be worth the counting. . . . Next time [we land on Mars] we must ship out our bodies to stare close-up at the wonder that must be Mars. . . . Mars is a dead world waiting to be stirred awake. . . . Some say we cannot afford the expense. We can't afford not to."

With these words, Ray Bradbury captures the essence of human response to the vast universe around us: a feeling of aloneness, coupled with a desire to connect to the cosmos; a yearning to learn what exists, and then to visit these faraway orbs; and a belief that we humans can bring life to dead worlds. These deeply ingrained notions resonate with our most profound feelings about the meaning of life. They deserve respect from those who organize our efforts in cosmic exploration, both for their spiritual importance and for their political—in the broadest sense—impact.

But although I share many of these feelings, there are strong arguments against sending humans to Mars, at least until several generations of automated spacecraft have taught us more about our celestial neighbor.

What are the arguments against sending humans to Mars? Intuitively, none exist: The difficulty, the danger, the expense all represent obstacles to be overcome in order to achieve the immense reward of exploring another planet not with robots but with real human beings, capable of shifting their attention and location in response to what they find, sensitive to the triumph and

glory of setting foot on a new world—"face to face . . .
with something commensurate to [the human] capacity
for wonder," to quote F. Scott Fitzgerald's ending to *The
Great Gatsby*. No one should overlook the driving power
of human desires: If the human race survives another
few centuries, I think it is a near-certainty that humans
will personally explore Mars, and the sun's other planets
as well. In the shorter run, however, the argument seems
persuasive (at least to me!) that human exploration of
Mars costs so much, and involves such large risks of contaminating another planet with Earth life, that we
should concentrate on a systematic examination of the
planet with robotic explorers.

When we cast a cold, dare I say scientific, eye on in-
person planetary exploration, I think it is evident that it
suffers from the same drawbacks that a British noble-
man was said to have found in the sexual experience:
"The pleasure is transitory; the expenses are enormous;
the position, ridiculous." It is important to remind ourselves that cold-minded analysis does not and should not
govern all our activities, and that human exploration of
Mars may offer such psychic benefits that we finally
choose to achieve it despite what logic may tell us. Nevertheless, the fact remains that so far as human trips to
Mars are concerned, the expenses *are* enormous, at least
ten and quite probably a hundred to a thousand times
greater than those involving automated spacecraft; the
pleasure *is* transitory, since we can receive the crucial
psychic reward, of knowing that we have reached Mars
in person, only once; and the position *is* ridiculous, since
logic tells us that the time to send humans to Mars will
arise once we have acquired sufficient knowledge from

automated spacecraft to allow a well-executed plan for human exploration to occur. In 1961, President John Kennedy struck a responsive chord among his fellow Americans by announcing that "we *choose* to go to the moon." I think a good case can be made that the rush to send humans to the moon not only had its origin in arms competition with the Soviet Union but also led to the absence of human explorers on the moon over the past twenty-five years. Lacking a coherent rationale, we found lunar exploration not "worth" pursuing.

In September 1996, five weeks after the first announcement of possible ancient life on Mars, President Clinton stated that for the foreseeable future, the United States' exploration of Mars would be carried out by sending instrument packages, not spacecraft carrying human beings. The Clinton administration's study concluded that the goal previously established by the Bush administration, to send humans to Mars (and arrange to bring them back!) by the year 2019, was too costly and too risky to be supported. At the same time, Clinton called for a meeting of experts in December 1996 to plan how best to explore Mars with automated spacecraft.

The President had timed his statement to coincide with the conclusion of the longest spaceflight by an American, the astronaut Shannon Lucid, who spent six months with two Russian scientists in the Mir spacecraft. Lucid's impressive feat, second in duration only to a yearlong space mission by a Russian astronaut, helps to show that humans can survive weightlessness, as well as the other disorienting stimuli (or lack of stimuli) that are associated with spaceflight, for lengths of time similar to those required for a journey from the Earth to Mars. The problem with sending humans on these journeys lies not with whether they can bear the ride, but rather with whether we can bear the expense for the modest returns that can be expected from the first human visits to Mars, and are willing to risk contaminating another planet with our own microbes, so that we never know whether any life forms found on Mars are endemic to that planet or are escaped Earth life.

In one sense, no conflict exists between the urge to explore Mars with robots and the desire for a human presence there; everyone recognizes that the first of these must precede the second, and that several successive automated missions will be needed to prepare for human activity on Mars. On the other hand, we do have an essential conflict between our hearts and our minds: The former want to put people on Mars, while the latter want to learn about conditions there. On the third hand, just as we have learned to adopt new attitudes concerning our relationship with Earth's environment, we can and probably shall achieve new views of how we relate to the cosmos. "You have to respect the life-forms that may exist [on Mars]," says the astronomer Jill Tarter. "If life exists there, it's not the same situation as if the planet were uninhabitable." This attitude may yet gain sufficient adherents to slow plans to send humans to Mars, if the fact of limited resources doesn't do the job on its own.

I strongly believe that we can learn to see automated spacecraft as extensions of ourselves. When astronauts took the first picture of the whole Earth from space, what most thrilled humanity was the fact that people could see their entire planet for the first time, and not that humans had gone into space to make that possible.

Future Missions to Mars

If robotic explorers find fossils on Mars, I think this news could be just as exciting as if human explorers had done so. The excitement, after all, lies within each of us. Of course, we respond most readily to other humans, but as we achieve a greater awareness of our place in the cosmos, I predict that our response will be more to care about what we learn than about who or what teaches us.

The Far Future and the Terraforming of Mars

In the interest of equal access, let me freely anticipate that sooner or later humans will indeed land on Mars, taking care (I hope!) not to disturb the environment. There they will establish scientific outposts from which to survey the sun's fourth planet. In science-fiction novels, this marks the beginning rather than the end of the story, and indeed we may anticipate that Mars will acquire a continuous human presence at some time during the next millennium. What can we anticipate from this presence and activity?

Some scientists, including Chris McKay, are intrigued by the possibility of *terraforming* the planet—altering it to resemble Earth more closely. "There are two broad concepts when it comes to terraforming Mars," says McKay. "One is that we could make Mars something like Earth was before our air grew rich in oxygen. That would mean making Mars's atmosphere much thicker, but still mainly carbon dioxide. To do this, we'd have to warm Mars up until the polar caps sublimated and put their carbon dioxide in the atmosphere." The addition of large amounts of carbon dioxide would allow the Martian atmosphere to produce a significant greenhouse effect, which would trap solar heat, prevent the gaseous carbon dioxide from solidifying again at the poles, allow liquid water to exist, and keep the entire Martian surface several tens of degrees warmer than it is now. And how could we melt the polar caps of Mars? "Supergreenhouse gases," says McKay. "CFCs [chlorofluorocarbons]—the stuff that's bad for Earth because we don't want to make our planet any warmer. They're tremendously more efficient than ordinary greenhouse gases like carbon dioxide." A relatively small amount of these supergreenhouse gases, introduced into the Martian atmosphere, could warm the planet by tens of degrees for many years, until they eventually disappeared, from the atmosphere, after many thousands of years, by combining with other molecules at high altitudes.

And what about making a carbon-dioxide atmosphere into an oxygen-rich one? "The second step, to make Mars really like the Earth, is much harder. You can evaporate the polar caps with a relatively modest amount of energy, but to get oxygen into the atmosphere, you have to break the chemical bonds that now hold oxygen in the soil. To do that, you need self-replicating machines called plants." In McKay's vision of step two, the terraforming of Mars would proceed through countless algaelike plants, which could live in the Martian lakes produced by step one. Like plants on Earth, the algae would take in carbon dioxide and release oxygen. "Melting the polar caps could take a few decades," he says, "but step two would require something like a hundred thousand years. You can't speed up the efficiency with which plants turn carbon dioxide into oxygen. But eventually you could have a global biosphere on Mars as rich and productive as the one on Earth."

Of course, not everyone sees the terraforming of Mars as a desirable goal. "I think it's deeply irresponsible to suggest that colonizing another planet will solve our problems with pollution and overpopulation," says Tobias Owen. "We'll just carry those problems with us. We should concentrate on terraforming the *Earth*— converting our planet into a more desirable place to live."

Chris McKay tends to look beyond these concerns. "I see the issue as divided into three parts," he says. "Can we? Should we? Will we? As to 'Can we,' I think there is no doubt that we can do so eventually. Should we? My answer is yes, and is based on a preference for life. Going out and spreading life is an important and useful thing to do." This is the attitude that Crick and Orgel assumed might well be present in extraterrestrial civilizations considering the desirability of directed panspermia. "And will we?" McKay asks. "I think not. It costs a lot—and for what return? Of course, it will be the humans on Mars, not those on Earth, who will be the main beneficiaries, and who will therefore be willing to donate a large fraction of their GNP to this project."

To me, terraforming Mars resembles the notion of creating a theme park at the summit of Mount Everest. It may happen, but we shall lose something valuable in the process; how much more, then, should we shun the opportunity to redo an entire planet? As a good citizen, I am ready to put my faith in the people, and in the leaders they choose democratically; the latter, as Bertrand Russell pointed out, cannot be less intelligent than the former. From this firm foundation I look confidently to the future. It is time to end our survey of what the rock from Mars had to tell us with a look at how it fits into the framework of science, along with a survey of some of the thoughts that the news from ALH 84001 caused to ripple through our psyches.

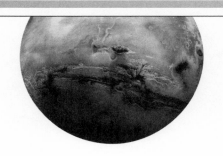

"Who are

you going

to believe—

me or your

own eyes?"

SCIENCE
AS A WAY
OF LIFE

The tale I have told of the search for Martian life and the search's implications for life elsewhere in the universe folds into a larger story. ALH 84001 and its ramifications provide an excellent example of how science advances as the result of hard work, intelligent insight, and experienced speculation. Most vividly and firmly, the discoveries and discussions concerning possible ancient life on Mars help to delineate the difference between the attitudes with which most people view the world and the ways that scientists explore the cosmos.

The Differences Between Scientific and Other Worldviews

Science is organized skepticism. Both the adjective and the noun that follows it play key roles in defining how scientists view the world—when they are thinking as scientists. Skepticism allows scientists, and anyone else who employs it, to doubt what they are told. It may seem odd that doubt lies at the root of science, which seems to consist of a host of facts and formulae that most of us must take on faith, and that even then lie only at the edge of comprehension. However, every scientist in training learns countless stories that highlight the importance of skepticism by describing men and women who doubted the conventional wisdom and broke through to a new understanding.

Organization allows the skepticism of scientists to advance the cause of knowledge. When a scientist announces a new breakthrough, the first reaction of that scientist's colleagues is a highly organized one: They doubt it. They do so because they know full well, from both experience and verified anecdotes, that most new ideas prove to be wrong. Science has long been organized to reward not only the creation but also the rejection of new ideas. This organization creates a hierarchy of skeptics, one that roughly reflects the significance of their doubts. Most important are the established experts in the same field of scientific endeavor—experts who achieved their status by advancing results that passed the doubt test. Next most important are others who work in the field, striving to advance their reputations and thus their careers. An excellent way to do so, just as good or even better than producing a new result on one's own, is to spot a fatal flaw in another scientist's bold new hypothesis.

This may seem a harsh way to run an enterprise, and certainly goes against what we consider good behavior in personal interactions. Scientists engage in it, and make it their bedrock principle, because they find it by far the most effective way to increase understanding. Because skepticism rules the world of science, departures from its principles cannot long survive the attacks of the doubters. If a scientist were to adopt a theory because it feels good, or satisfies the boss's desires, or brings additional grant money, any temporary gain that might result would be more than offset by the eventual conclusion that this scientist was in error. Although some errors can be glorious, and although a carefully conceived hypothesis that later proves erroneous will bring no shame to its creator, the highest rewards go to those theories that prove to be both significant and correct.

Within the scientific arena, each scientist competes for his or her colleagues' attention by citing the results of his or her endeavor and waiting for the organized, skeptical reaction. Those who hear these results promptly ask, Why should I think this is right? This begins the testing process that yields either rejection or acceptance of the new statement about nature; if it passes the tests, scientists will call it true (though it will always be subject to possible correction by later discoveries and analyses). Of course, scientists do not spend their days full of doubt about everything they hear or read. Instead, they save their skepticism for what is new and important. Their reward comes from the fact that when they believe something, they do so in a different way; their belief rests on having overcome, by direct testing, their own and (even more important) their colleagues' skepticism.

The famous physicist Richard Feynman expressed this well when he dismissed assertions that ancient astronauts had visited the Earth by stating that he knew how hard he had to work to show that something is true. Feynman did not mean that ancient astronauts could not have visited the Earth, only that those claiming this to be so had utterly failed to participate in the scientific testing process. Why, then, did Feynman not investigate this highly important assertion for himself? Because he clearly judged such an investigation unlikely to yield useful results; in short, he had more important things to do. Like Abraham Lincoln's studied refusal to admit Jefferson Davis's existence during the Civil War, scientists' unwillingness to pay attention to ideas that arise outside the community of scientists is maddening to some of those outsiders—for example, to those who have insisted that the "Face on Mars" cries out for further study. This refusal forms part of the *organized* skepticism of science, which has in effect created a "union shop." Those who have not joined the organization get little attention or respect from scientists acting in their scientific capacities. Like scientific skepticism itself, this lack of respect may seem both harsh, elitist, and antidemocratic. So it is; but it has allowed science to succeed.

In addition to its nondemocratic attitudes, science also violates the most basic rule of human belief, that of giving credence to ideas that carry an emotional reward.

Science as a Way of Life

No one knows better than scientists that each of us believes what we like, and that reality need not conform to these beliefs. If science did not offer its practitioners the reward of perceiving physical reality to a degree not available in other belief systems, almost none of them would engage in activities that often challenge their inner feelings about the cosmos. Even with all their attempts to suspend their emotions and prejudices, scientists recognize their dangers, leading to the scientific maxim that a theory cannot be completely refuted, but may eventually die when its originators succumb to mortality. With organized skepticism, science matches one person's set of prejudices with another's different set (so it is hoped!). The debate over reality then has a good chance to point in the proper direction.

Note that the organized skepticism of science requires that scientists must compete for their colleagues' attention. Scientific research creates many competing demands on a limited amount of time, and its rewards accrue only by having a sizable or an influential fraction of one's colleagues (or even better, both) take note of one's efforts. In this regard, nothing succeeds like a great track record. A highly successful scientist once told me of his experience as a newly made Ph.D., when he attended a scientific conference and found to his dismay that although he had published an article dealing with the exact subject under discussion, no one seemed to have read it. "I took a vow on the spot that that would never happen again," he said. Not surprisingly, this man, a member of the prestigious National Academy of Sciences, has achieved a large amount of notice, which would be worthless to him or other scientists—or to the world—

unless what he had to say had survived the doubts of his peers.

Because the scientific worldview differs so markedly from that of most people, scientists often meet psychological barriers when they attempt to explain why they believe certain statements to be correct, and even more so when they try to show why completely wacky theories are just plain wrong. Aware at a deep level that organized skepticism will never be accepted by most of humanity, they often fall into a common human error and adopt the attitude well expressed by the author Ring Lardner about his father: " 'Shut up,' he explained."

In the discussion above, I have considered science in its noblest form. In the real world, some scientists do adopt theories, and select lines of research, because they feel good, or satisfy the boss, or bring in more money. Scientists, being entirely human, remain subject to the frailties of the human condition. They often make mistakes, both scientific and personal, including a failure to observe the conventions of scientific debate. Attempts to demonstrate that another scientist has fallen into error can be amazingly intense without bringing opprobrium (and indeed will often attract the admiration of fellow scientists), whereas attacks on a scientist's personality or behavior cross the same sort of line that we observe in daily living. The first type of activity advances scientists' attempts to discover the truth about the world, while the latter simply adds to our total of personal anguish.

The appearance of the word "truth" reminds us that some nonscientific academics, themselves seeking to advance theories that their colleagues will accept, have promoted the notion that the truth about nature cannot be found and that instead, each of us constructs his or her own "truth," some of which society marks as "the" truth. The denial that nature has actual reality that we can discover, some of it in practice and the rest in theory, seems to most scientists too ridiculous to discuss; they lump it together with proclamations of the "end of science" (roughly speaking, the notion that scientists have run out of testable hypotheses). One of the grave flaws of the reality-as-human-construct hypothesis is that we cannot test whether or not this theory is correct. Of course, this objection weighs heavily only among scientists, who may be said to be prejudiced in this debate.

The Mars Meteorite and the Scientific Worldview

With ALH 84001, science was clicking on all cylinders. Each of the fields involved in the analysis of the Mars rock—geology, astronomy, chemistry, microbiology, paleontology, molecular biology, physics, and materials science—contributed both to securing the data and to debating their significance. What remains most striking about the meteorite from Mars is that its contents, though highly provocative, remain a mystery, in the fundamental sense that agreement has not yet arisen about their nature and implications. Nothing could be more characteristic of science. During the next few years, as scientists compete for precious samples of ALH 84001, new Martian meteorites are identified, and spacecraft return new data from Mars, we may be able to resolve what the evidence in the meteorite can tell us. If not, scientists will have to try harder to secure still more evidence and to debate its significance, following the path of organized skepticism that has guided them so well.

The power of organized skepticism becomes especially evident when we look back at a phenomenon such as the furor over cold fusion that arose in 1989. Unlike the evidence in the rock from Mars, which skeptics could quickly verify as real even as they argued about its implications, the evidence alleged to demonstrate cold fusion quickly evaporated as other scientists strove in vain to confirm it. If science had not been organized in a way that made this rapid response possible, the cold-fusion excitement would have persisted for years, consuming large amounts of resources and devouring scientific careers. Cold fusion would have been wonderful—had it been true. But not all the wishing in the world could make it true, and we are all better off for the speed with which cold fusion was disproven.

At the risk of repetition, let us also note that science offers one particular way to regard the world, a view that has yielded great benefits not only to its adherents but also to those who know nothing of science. However, you never gain something but that you lose something, as Thoreau said. Since science depends on its skepticism, an attitude different from what we feel in our inner selves, we should never lose sight of the fact that science is only one way to tell a story, only one myth that "explains" the world around us. Science differs only because the story and myth concern what is generally called physical reality; for those who find this fact incomplete or even low on the hierarchy of importance, science can never rank as much more than a source of fascinating tidbits of information. Many a scientist feels the split between his or her skeptical side, which leads to greater understanding of the physical world, and a more ethereal side, which takes an interest in other matters. In what may be seen as an attempt to unite these two aspects, Albert Einstein once said that "the most incomprehensible thing about the world is that it is comprehensible." A fair amount of mystery still lies within that word "comprehensible."

Theological Implications of Possible Ancient Life on Mars

Among the many interesting reactions to the news from ALH 84001, none stood out more than those of theologians of various denominations. The possibility of life on Mars—even ancient life one-millionth of an inch across—raises once again the question of humanity's place and purpose in the universe. Astronomers, who had once dethroned the Earth as the core of creation by showing that the sun, not the Earth, forms the center of the solar system, now seemed to be at it again, this time with life from another planet. To be sure, tiny fossils hardly posed the question as strikingly as the discovery of alien intelligence would have. John Cobb, an emeritus professor at the Claremont School of Theology in California, posed the theological problem this way: If "the one thing that might have seemed distinctive about [Earth]—that it provided the circumstances in which very complex beings came into being—if that's taken away from us, it would take us one more step away [from the idea] that humanity has some special role in creation."

In its organized theology, the Judeo-Christian tradition has more difficulty accepting the notion of extraterrestrial life than most other major religions. The Koran, the holy book of Islam, uses the word "universes" and makes it clear that the cosmos includes creations that humans cannot know. Nor do Buddhism or Hinduism experience difficulties with the concept of life, even highly developed and intelligent life, existing elsewhere in the universe; they may even be said to welcome it. But the Bible and its interpretation do cause problems. Though we risk oversimplification, we may divide the Judeo-Christian response to extraterrestrial life into two basic categories, the Earth-centered and the greater-glory camps. The latter group sees life elsewhere in the universe as just another jewel in the crown of creation. For theologians in this category, the story of Genesis simply omits what may have happened on other worlds. In response to the news from Mars, the spokesman for the Catholic archbishops of France, Olivier de la Brosse, said, "Christian theology has never supposed that only Earth and its inhabitants are the works of God. The Bible deals only with the history of humanity. If it is confirmed that life exists on Mars, for me this is not a supplementary proof but a manifestation of God's power. I conclude that God is still greater than I thought."

Like de la Brosse, many Protestants conclude that various worlds could have a claim to heaven. Nancy Murphy, who teaches Christian philosophy at the Fuller Theological Seminary in Pasadena, California, responded to the news from Mars by stating, "God could reveal himself to other beings in appropriate form." (Actually, most definitions of God allow Him to reveal Himself even in inappropriate form.) The greater-glory camp may have a more difficult problem with the existence, if it were to be proven, of ancient life on Mars that later became extinct. "If one puts that into a theological framework," says Philip Hefner of the Chicago Center for Religion and Science, "does this mean that God's experiment with life sometimes fails and sometimes doesn't?" For individual species on Earth, the answer seems clear, but here we might be talking about an entire planet. Nevertheless, Claremont's John Cobb says, "for most theologians today, this discovery will have no effect whatsoever." (Let us hope that Cobb is referring to theologians' religious beliefs rather than their overall understanding of the cosmos.) "But that's very different," Cobb continues, "from popular religious belief, [where] what people learn from science has an effect on how they think religiously."

This brings us to the second category, the Earth-centered system of belief. At an intuitive level, everyone on Earth belongs to this group. Many of us, however, have learned, and have accepted at some level in our inner selves, that the sun is one star among several hundred billion in the Milky Way galaxy, and that our galaxy is one among hundreds of billions, or even more, in the visible universe. Such thoughts can be unsettling; they certainly do not conform to what our brains perceive. As a result, many cultures have experienced sad episodes of attempting to repress those who spread the news about the small Earth and mighty cosmos. These attempts arise from an understandable, but hardly commendable, desire to keep society on an even keel by assuring a continuity of belief in space and time. When the Catholic Inquisition burnt Giordano Bruno at the stake in 1600 for proclaiming that other worlds have inhabitants whom God loves as much as he loves humanity, it believed it was doing a good deed for society, even though it was in fact making it more difficult to recognize the glory of God. (Too bad that Olivier de la Brosse was not present to explain to his fellow Catholics that the Bible refers only to the story of humanity!) All theocracies have eventually encountered similar problems with "heresy," since all groups of humans have exhibited a delightful difficulty in believing the same thing for long periods of time.

At the individual level, today we are free to believe what we choose. For some Catholics and Jews, and for larger numbers of fundamentalist and evangelical Protestants, this includes a so-called literal interpretation of the Bible, a task at least as a difficult as a literal interpretation of the United States Constitution. For many fundamentalist Christians, the words in this book deserve no respect, since they contradict the word of God. Evolution is a hoax, they say, and Earth is the only planet with intelligent life. "Our theology would allow for simple life-forms but not for other thinking beings on a par with or superior to humanity," says John McArthur, the president of Master's College in Santa Clarita, California. In that theology, God created the world in six days, and each of those days contained twenty-four hours.

It might seem that a theology capable of dismissing the evidence for evolution of thousands of species on Earth could hardly be fazed by possible microorganisms in a rock from Mars. However, there is something about the Martian mystique, even in fundamentalist circles, that seems to raise the evolution issue anew. Perhaps this is the feeling that after millions of fossils on Earth had to be rejected as misinterpreted or bogus, suddenly the daunting prospect of a slew of fossils from another planet appears. In September 1996, when David McKay, Richard Zare, and Wesley Huntress, Jr., testified before the Space Subcommittee of the House of Representatives' Committee on Science, the first question from Ralph Hall, a congressman from Texas and an important member of the subcommittee, dealt with the 4.5-billion-year age of ALH 84001. The number of years involved was hard for people to reconcile with the biblical account, Hall said. In their responses, the scientists refrained from either lecturing Hall or offering ways to square things by, for example, positing that God created the world complete with an ancient record in the rocks, perhaps to keep us on our toes. (This compromise between fundamentalism and evolutionary theory, basically untestable and unappealing to those on both sides of the issue, deserves praise for its bold attempt to seize and hold the middle ground.) Instead, the scientists stressed the benefits of scientific research and expressed their hope that further investigation would reveal many more facts about the cosmos.

The New Planets

By a happy coincidence, 1996 brought not only news of possible life on Mars but also of a slew of new planets orbiting stars near the sun. Using advanced detectors and computer systems, astronomers in Europe and the United States discovered the first confirmed planets that orbit sunlike stars, not by observing the planets themselves but rather by detecting the effects that the planets' gravitational forces produce on their stars. The first extrasolar planet, orbiting the star 51 Pegasi, was discovered with this method in October 1995. One year later, the number of planets found around other stars had climbed to eight, with every indication of going significantly higher in a few years' time. Suddenly, after decades of frustration in the attempt to find astronomically tiny planets nestling relatively close to stars, the dam had burst. Observations demonstrated convincingly that a sizable fraction of stars like the sun have planets, a fact that astronomers had long suspected but lacked the evidence to prove.

The combination of planets orbiting sunlike stars plus indications of ancient life in a rock from Mars might appear to have dealt the Earth-centered cosmos a fatal evidentiary blow. However, as we have noted in earlier chapters, all evidence remains subject to interpretation. If we like, we can regard the belief that Earth stands higher than all other worlds in God's eye as a manifestation of the human spirit, itself a product of whatever powers created the universe and its worlds. But the scientific attitude cannot refrain from asking, Why should I believe this? What evidence exists for the notion that we on Earth are the acme of creation? To which theologians may reply, You are asking the wrong question.

Extraterrestrial Life Throughout the Universe?

One result of the discovery of new planets and possible ancient life on Mars seems particularly striking: The likelihood of extraterrestrial life has increased—not in reality, which remains as it was before the recent discoveries, but in our estimates of the probabilities. Before late 1995, astronomers felt sure that sunlike stars must have planets, as nothing about our sun distinguishes it from its many thousands of near-twins in the Milky Way. On the other hand, no one had *detected* any of these putative planets, a fact that could be explained by the technical difficulties involved but that was nonetheless troubling. Finding the new planets has lifted a mental burden, leaving astronomers one step closer to obtaining proof that life, and perhaps intelligent life at that, is widespread in the universe.

The news from Mars likewise cheers scientists who believe that Earth-like conditions elsewhere in the universe should have produced, in most if not all cases, other forms of life that inhabit worlds other than ours. Although ALH 84001 does not contain definitive proof of life on Mars, this ancient Martian rock does seem to make more likely the possibility that Mars once teemed with life. Should this turn out to be true, theories that imply that life has arisen on planet after planet will receive validation. If two neighboring planets in the same system possess histories that include life, the argument that life arises only rarely, and under highly specific conditions, vanishes into thin air. One might attempt to resurrect it by invoking panspermia, but this would only prove that even though the actual origin of life may be a rare event, life spreads relatively easily from place to place in the cosmos. Most scientists—and most of the general public—would consider other forms of life quite fascinating even if they evolved from cosmic seeding. In particular, we would be highly interested in whether extraterrestrial forms of life, no matter how they originated, have evolved to our stage of development, or even beyond.

On this point we remain almost as ignorant as we were before the discoveries of extrasolar planets and possible ancient life on Mars. If we increase our estimates of the number of worlds that are favorable to life, and of the number of those worlds on which life has appeared, this might seem to require an increase in our estimate of the number of places that might have given rise to a society capable of, and interested in, communication across interstellar distances.

On the other hand, we have no good evidence that any such societies exist. The evidence adduced in favor of extraterrestrial visitors to Earth consists either of individual memories and observations, or of physical evidence similar to the Face on Mars, easily seen to be far from anything that would convince a reasonable jury. Although eyewitness testimony is admissible in most courtroom proceedings, one can argue that when dealing with strange events, such accounts have proven so unreliable that they cannot be trusted. The weirder the alleged occurrence, the less one can rely on eyewitnesses—a fact well known to psychologists, though anathema to the human psyche. In the movie *Duck Soup*, Chico Marx asks the rich lady, "Who are you going to believe—me or your own eyes?" And he is right to do so, at least in this case, for he is telling the truth: It was not Chico but Harpo whom the lady saw. From a scientific viewpoint, you can believe neither me nor your own eyes. Until we have hard physical evidence of extraterrestrial visits—

something a good deal more convincing than, for example, crop circles that hoaxers can easily make—we should reject amazing hypotheses in favor of more mundane ones.

The views above are, of course, entirely my own (unless they could lead to trouble with believers in a conspiracy to suppress UFO information, in which case I have no idea whose they are). From a scientific viewpoint, estimating the existence of highly advanced civilizations elsewhere in the cosmos rests about where extrasolar planets did some years ago: On the basis of the large numbers of possible sites for life, scientists think advanced civilizations *should* exist, but have no convincing evidence that they do. The *absence* of documented extraterrestrial visitors to Earth, along with the failure to detect any radio signals from extraterrestrial civilizations, argues in a modest way against the existence of other civilizations, at least in our corner of the Milky Way galaxy.

However, we should remember that the same fact that argues in favor of a large number of sites for life— the enormous volumes of space in the universe, which contains huge numbers of stars with, presumably, their own planets as well—also argues against interstellar travel. Interstellar distances dwarf those in our solar system: The closest stars to the sun are a million times farther from us than Mars at its closest. If we make a model of the Milky Way in which stars have the sizes of light bulbs, each of these lights would have a large city to itself, with one bulb each in New York, Boston, London, Munich, Athens, and Delhi. Although we cannot predict how rich other civilizations may be in energy terms, physical realities suggest that they would start their attempts to contact any neighbors not with incredibly long journeys in person, but rather with radio signals. In that case, humanity owes it to itself to increase the relatively modest efforts that are now under way to listen, so far without success, for possible signals from intelligent forms of life around other stars.

From the discovery of a piece of rock on ice to the quest for other civilizations may seem a large step, but when the rock comes from another planet, was formed more than four billion years ago, and may contain material formed by ancient extraterrestrial life, the leap seems less gigantic. Only time can tell whether we shall look back upon 1996 as the year in which the first extraterrestrial life was discovered, billions of years after it lived, millions of years after its remnants left its planet, and thousands of years after those remnants fell to Earth.

Whatever the resolution of its enigmatic contents, ALH 84001 has, I hope, furnished my readers with a glorious ride through the cosmic story of life, so far as we know it today. With any luck, the time will soon come when we not only comprehend what the meteorite from Mars has to tell us, but also understand the origin and development of life in the solar system. This history awaits our discovery—if we can create the instruments that will "prospect" for life, send them to explore places where past or present life seems reasonable, and use them to analyze their discoveries with an organized skepticism that will allow us to discern the truth about life in the universe. Thanks for traveling with me on part of the journey; the rest is spread out before you.

GLOSSARY

Abiotic—Without the presence of life.

Accretion—An infall of matter that adds to the mass of an object.

Amino acid—A class of relatively small molecules, made of thirteen to twenty-seven atoms of carbon, nitrogen, hydrogen, oxygen, and sulfur, which can link together to form proteins.

Archaea (singular Archaeon)—Representatives of one of the three domains of life, thought to be the oldest type of life on Earth. All Archaea are single-celled and thermophilic (capable of thriving at temperatures above 50–70 degrees Celsius).

Asteroid—One of the small objects, made mainly of rock or of rock and metal, that orbit the sun, mainly between the orbits of Mars and Jupiter, and range in size from six hundred miles in diameter down to objects less than a few hundred yards across.

Astronomical Unit—The average distance from the Earth to the sun, equal to 149,597,900 kilometers, or 92,955,000 miles, and abbreviated as A.U.

Atom—The smallest electrically neutral unit of an element, consisting of a nucleus made of one or more protons and zero or more neutrons, around which orbit a number of electrons equal to the number of protons in the nucleus. This number determines the chemical characteristics of the element.

Bacteria—One of the three great domains of life on Earth, formerly known as prokaryotes, single-celled organ-

Glossary

isms with no well-defined nucleus that holds the cell's genetic material.

Basalt—A type of volcanic rock, highly abundant in the Earth's crust.

Biomagnetite—Magnetite (a magnetic mineral) that has been produced or significantly altered by living creatures.

Biomineral—A mineral important to certain classes of living organisms.

Biosphere—The totality of all living matter on Earth, including life in and on the atmosphere, oceans, streams, lakes, subsurface, and land surface.

Carbohydrate—A molecule made only of carbon, hydrogen, and oxygen atoms, typically with twice as many hydrogen as oxygen atoms.

Carbon—The element made of atoms whose nuclei each have six protons, and whose different isotopes each have six, seven, or eight neutrons. The two long-lasting isotopes of carbon have either six neutrons (99 percent) or seven neutrons (1 percent) per atom.

Carbonaceous chondrite—A member of the class of the oldest, least altered meteorites, characterized by carbon-rich inclusions called chondrules.

Carbonate—A class of minerals made of carbon and oxygen atoms, together with other types of atoms that include calcium, magnesium, iron, manganese, sodium, and zinc, in which each carbon atom binds to three oxygen atoms and the resulting carbon-oxygen assembly binds to other atoms.

Carbon dioxide—A type of molecule containing one carbon and two oxygen atoms, whose chemical symbol is CO_2.

Catalyst—A substance that increases the rate at which certain reactions between molecules occur, without being itself consumed in these reactions.

Cell—A structural and functional unit found in all forms of life on Earth. Earth life ranges in size from the smallest single-celled organisms up to plants and animals that each contain many thousand trillion cells.

Celsius (Centigrade) temperature scale—A scale of temperature that registers the freezing point of water at 0 and its boiling point at 100 degrees.

Chert—A type of sedimentary rock, made up primarily of microscopic silica crystals.

Chondrite—A meteorite that contains inclusions called chondrules.

Chondrule—A small, round granule of matter embedded within some meteorites.

Chromosome—A single DNA molecule, together with the proteins associated with that molecule, which stores genetic information in subunits called genes and can transmit that information when cells replicate.

Comet—A fragment of primitive solar-system material, a "dirty snowball" made up of ice, rock, dust, and frozen carbon dioxide ("dry ice"), typically with an orbit around the sun much larger than any planet's.

Compound—A homogeneous substance made of atoms of two or more different elements in constant proportions—that is, of a particular type of molecule.

Cosmic rays—Particles moving through interplanetary and interstellar space at nearly the speed of light, most of which are proton, electrons, and helium nuclei.

Cosmic seeding—An alternative name for panspermia.

Daughter nucleus—A nucleus that forms from the radioactive decay of another type of nucleus, called the parent.

Directed panspermia—The concept that extraterrestrial civilizations could have sent various forms of life to colonize planets such as our own.

DNA (deoxyribonucleic acid)—A long, complex molecule, consisting of two spiral strands, bound together by thousands of cross-links formed from small molecules. When DNA molecules divide (replicate), they split lengthwise along their cross-linking molecules; each

Glossary

half of the molecule can then re-form a complete molecule of DNA from smaller molecules that exist in the
nearby environment.

Domain—One of the three main types into which Earth life
is now classified: Archaea, prokaryotes, and eukaryotes.

Dry ice—Frozen carbon dioxide (CO_2).

Earth life—The totality of all life on Earth, characterized
by a chemistry based on carbon atoms and the use of
water as a solvent. This term is not yet in general use,
but is employed in this book for clarity.

Electron microscope—A microscope that produces en-
larged images either by bouncing electrons off a substance (scanning electron microscope) or by shooting
electrons through it (transmission electron microscope). The effects that the substance produces on the
electrons reveal the details within it.

Electron—An elementary particle with one unit of negative
charge, which orbits the nucleus in an atom.

Element—The set of all atomic nuclei that have the same
number of protons in the nucleus.

Enzyme—A type of molecule, either a protein or RNA, that
serves as a site at which molecules can interact in certain ways, and thus acts as a catalyst, increasing the
rate at which particular molecular reactions occur.

Eukaryote—An organism, either single- or multicelled,
that keeps the genetic material in each of its cells within
a membrane-bounded nucleus.

Europa—One of Jupiter's four large satellites, intriguing
for its covering of water ice, which may conceal liquid
water.

Evolution—In biology, the result of the process of "natural
selection" (differential success at reproduction), which
under certain circumstances causes groups of similar
organisms, called species, to change over time so that
their descendants differ significantly in structure and
appearance.

Glossary

Evolutionary distance—For any pair of organisms, a measure of the similarity of the organisms' gene sequences; small evolutionary distances imply a greater degree of similarity.

Extrasolar—Pertaining to objects beyond the solar system; extrasolar planets are planets that orbit stars other than the sun.

Face on Mars—A mile-wide rock outcropping on the Martian surface that reminds some observers of a human face.

Fossil—A remnant or trace of an ancient organism.

Frequency—Of photons, the number of oscillations per second.

Galaxy—A large group of stars, typically numbering in the hundreds of millions up to the hundreds of billions, and usually containing significant amounts of gas and dust, held together by the mutual gravitational attraction among the stars.

Galilean satellites—Jupiter's four largest satellites, discovered in 1610 by Galileo.

***Galileo* spacecraft**—The spacecraft sent by NASA to Jupiter in 1990, which arrived in December 1995, dropped a probe into Jupiter's atmosphere, and continued to orbit the giant planet, photographing it and its Galilean satellites.

Gene—A section of a chromosome that specifies, by means of the genetic code, the formation of a particular chain of amino acids.

Genetic code—The set of "letters" in DNA or RNA molecules, each of which specifies a particular amino acid and consists of three successive molecules like those that form the cross-links between the twin spirals of DNA molecules.

Genome—The total complement of an organism's genes.

Geochemistry—The chemistry of the composition and changes in the Earth's crust.

Giant planet—A planet similar to Jupiter, Saturn, Uranus, or Neptune, consisting of a solid core of rock and ice surrounded by hydrogen and helium gas, with a mass ranging from a dozen or so Earths up to many hundred times the mass of Earth.

Greenhouse effect—The trapping of infrared radiation by a planet's atmosphere, which raises the temperature on and immediately above the planet's surface.

Greigite—An iron sulfide made of crystals whose units consist of three iron and four oxygen atoms.

Habitable zone—The region surrounding a star, a spherical shell bounded by inner and outer spherical surfaces, within which the star's heat can maintain one or more potential solvents in the liquid state.

Half-life—The time interval during which half of a particular type of radioactive nucleus will undergo radioactive decay.

Helium—The second lightest and second most abundant element, whose nuclei all contain two protons and either one or two neutrons.

Hydrocarbon—A molecule that consists entirely of hydrogen and carbon atoms.

Hydrogen—The lightest and most abundant element, whose nuclei all contain one proton and either no neutron or one neutron.

Hyperthermophile—An organism that thrives at temperatures close to the boiling point of water.

Igneous rock—Rock made from material that has been melted or partially melted, typically through volcanic action.

Infrared—Electromagnetic radiation consisting of photons whose wavelengths are all slightly longer, and whose frequencies are all slightly lower, than the photons that form visible light.

Inner planets—The sun's planets Mercury, Venus, Earth, and Mars, all of which are small, dense, and rocky in comparison with the giant planets.

Glossary

Inorganic—Not involving life or the chemistry on which life is based; in particular, not based on carbon atoms.

Interplanetary dust—Dust spread among the planets in the solar system and in other planetary systems.

Interstellar dust—Dust particles, each made of a million or so atoms, probably ejected into interstellar space from the atmospheres of highly rarefied stars.

Ion—An atom that has lost one or more of its electrons.

Iron oxide—A mineral made primarily of iron and oxygen atoms.

Iron sulfide—A mineral made primarily of iron and sulfur atoms.

Isotope—Nuclei that each contain the same number of protons, and hence belong to the same element, but contain different numbers of neutrons.

Kilogram—A basic unit of mass in the metric system, containing one thousand grams. On the Earth's surface, one kilogram has a weight of approximately 2.2 pounds.

Kilometer—A unit of length in the metric system, equal to one thousand meters and approximately 0.62137 mile.

Kinetic energy—Energy associated with motion.

Light—Photons whose frequencies and wavelengths fall within the band denoted as visible light, between infrared and ultraviolet.

Light-year—The distance that light travels in one year, equal to about six trillion miles.

Limestone—A type of sedimentary rock, consisting mainly of calcium carbonate, created underwater, often as the result of the accumulation of the shells of tiny marine organisms that manufacture calcium carbonate from the carbon dioxide and calcium in seawater.

Luminosity—Of an object, the total amount of energy emitted each second in photons of all types.

Magma—Melted rock formed within the Earth or Earth-like planets.

Magnetite—A form of iron oxide, made of groups of three

iron and four oxygen atoms, often with titanium or
magnesium atoms as part of the mixture.

Magnetotactic bacteria—Bacteria that create magnetic
mineral deposits and use the effects of Earth's magnetic
field on these minerals to orient themselves spatially.

Mass extinction—An event in the history of Earth life, in
some cases the result of a massive impact with Earth,
during which a significant fraction of all species of living organisms become extinct within a geologically
short period of time.

Metabolism—The totality of an organism's chemical processes.

Meteor—A luminous streak of light produced by the heating of a meteoroid as it passes through Earth's atmosphere.

Meteorite—A meteoroid that survives its passage through
Earth's atmosphere.

Meteoroid—An object of rock or metal, or a metal-rock
mixture, orbiting the sun, smaller than an asteroid or
planet.

Meteor shower—A large number of meteors, observed to
radiate from a particular point on the sky, the result of
the Earth's crossing the orbit of a meteor swarm.

Meteor swarm—A group of meteoroids that orbit the sun
along the same trajectory, with the density of the meteoroids varying at different points along the swarm.

Meter—The fundamental unit of length in the metric system, equal to approximately 39.37 inches.

Michelson-Morley experiment—A crucial experiment in
physics, first performed at the Case School of Applied
Science in Cleveland in 1887 by Albert Michelson and
Edward Morley, which demonstrated that the speed of
light through space is constant no matter what the motion of the source of light or of the light detector may
be.

Microbacteria (microbes)—Bacteria whose sizes are measured in microns.

Glossary

Micron—One-millionth of a meter, equal to 0.00003937 inch.

Microsecond—One-millionth of a second.

Milky Way—The galaxy that contains the sun and approximately 300 billion other stars.

Millibar—A unit of pressure, equal to one-thousandth of the average atmospheric pressure at Earth's surface.

Millimeter—One-thousandth of a meter, equal to 0.03937 inch.

Mineral—A naturally occurring solid that consists either of a single element or of a molecular compound, and that has a systematic internal arrangement of atoms with a definite chemical composition.

Molecule—A stable grouping of two or more atoms.

Monomer—A relatively small molecule that serves as one of the components of long-chain molecules (polymers). In Earth life, the best-known monomers are the amino acids that form protein molecules.

Murchison meteorite—A meteorite found near Murchison, Australia, in 1969, which contained organic molecules, including amino acids.

Mutation—A change in an organism's DNA that can be inherited (passed on from ancestors to descendants during replication). These changes create genetic diversity.

Nanofossil—A fossil whose size is measured in nanometers.

Nanometer—One-billionth of a meter, equal to 0.00000003937 inch.

Neutron—An elementary particle with no electric charge, stable when part of an atomic nucleus but subject to rapid decay when isolated.

Nitrogen—The element made up of atoms whose nuclei each have seven protons, and whose different isotopes each have six, seven, eight, nine, or ten neutrons. Most nitrogen atoms have seven neutrons as well as seven protons.

Nuclear fusion—The joining of two nuclei under the influence of strong forces, which occurs only if the nuclei approach one another at a distance approximately the size of a proton, about 0.4×10^{-13} inch.

Nucleic acid—Either DNA or RNA.

Nucleotide—One of the cross-linking molecules in DNA and RNA. In DNA, the four nucleotides are thymine, cytosine, guanine, and adenine; in RNA, uracil plays the role that thymine does in DNA.

Nucleus—(1) The central region of an atom, composed of one or more protons and zero or more neutrons. (2) The region with a eukaryotic cell that contains the cell's genetic material in the form of chromosomes.

Organelle—One of several types of specialized subregions within a eukaryotic cell.

Organic—Referring to chemical compounds containing carbon atoms as an important structural element; carbon-based molecules. Also, having properties associated with life.

Orgeuil meteorite—A meteorite found near Orgeuil, France, in 1964, which contained spores and other signs of life; these turned out to have entered the meteorite as contamination during its time on Earth.

Orthopyroxene—A type of silicate rock, primarily silicon and oxygen, plus some iron and magnesium, in its composition.

Oxide—A mineral containing oxygen ions bonded to metallic ions.

Oxygen—The element whose atomic nuclei each have eight protons, and whose different isotopes each have seven, eight, nine, ten, eleven, or twelve neutrons. Most oxygen atoms have eight neutrons to accompany their eight protons.

Ozone—Molecules made of three oxygen atoms (O_3). Ozone molecules high in Earth's atmosphere shield the surface against most ultraviolet radiation.

Panspermia—The hypothesis that life from one locale can

Glossary

be transferred to another, as from planet to planet in the solar system; also called cosmic seeding.

Parent nucleus—A nucleus that undergoes radioactive decay, producing a daughter nucleus.

Permafrost—Permanently frozen underground soil, overlain by a surface layer that thaws and refreezes each year.

Photosynthesis—The use of energy in the form of visible light or ultraviolet to produce carbohydrate molecules from carbon dioxide and water; in some organisms, hydrogen sulfide (H_2S) plays the role that water (H_2O) does in most of the photosynthesis on Earth.

Planet—An object in orbit around a star that is not another star and has a size at least as large as Pluto, the sun's smallest planet.

Planetesimal—An object much smaller than a planet, capable of building planets through numerous mutual collisions.

Plate tectonics—Slow motions of plates of the crust of Earth and similar planets.

Polymer—A long-chain molecule made of smaller molecules called monomers, linked together repetitively, with small but important variations.

Primitive atmosphere—The original atmosphere of a planet.

Primitive Earth—The Earth during the epochs lasting from 4.5 billion years ago, the time of the Earth's formation, through about 3.8 billion years ago, the end of the era of intense bombardment.

Prokaryote—One of the three domains of life, consisting of single-celled life in which the genetic material does not reside within a well-defined nucleus of the cell.

Protein—A type of large molecule made of one or more chains of amino acids.

Proton—An elementary particle with one unit of electric charge; one of the two basic components of an atomic nucleus.

Glossary

Protoplanet—A planet during its later formation stages.

Protoplanetary disk—The disk of gas and dust that surrounds a star, especially during the earliest part of the star's lifetime, from and within which planets may form.

Protostar—A star in formation, contracting from a much larger cloud of gas and dust under its self-gravitation.

Protosun—The sun during its formation process, which ended 4.5 billion years ago.

Pyrolitic—Pertaining to the chemical changes caused by heat.

Pyrrhotite—A type of iron sulfide, made from units of sulfur plus one or two iron atoms.

Radioactive decay—The process by which certain types of atomic nuclei spontaneously transform themselves into other types.

Radioactive nucleus—A nucleus capable of undergoing radioactive decay.

Radiometric dating—Establishing the ages of rocks by measuring the ratio of the number of nuclei formed by a particular type of radioactive decay ("daughter nuclei") to the number of nuclei ("parent nuclei") that produce the daughter nuclei when they undergo radioactive decay. The larger the ratio of daughter to parent nuclei, the longer the time that has elapsed since the rock formed.

Replication—The process by which a "parent" DNA molecule divides into two single strands, each of which forms a "daughter" molecule identical to the parent.

Ribosome—A complex of protein and RNA molecules, and the sites where proteins are assembled from smaller molecules.

RNA (ribonucleic acid)—A large, complex molecule, made of the same types of molecules that constitute DNA, which performs various important functions within living cells, including carrying the genetic messages

Glossary

embodied in DNA to the places where proteins are assembled.

Runaway greenhouse effect—A greenhouse effect that grows stronger as the heating of a planet's surface increases the rate of liquid evaporation, which in turn increases the greenhouse effect.

Satellite—A relatively small object that orbits a much larger and more massive one; more precisely, both objects orbit their common center of mass.

Sedimentary rock—Rock formed from material that has settled to the bottom of a liquid, typically from loose layers of sand, mud, minerals, and organic matter.

Self-gravitation—The gravitational force that parts of an object exert on all the other parts.

SETI—The search for extraterrestrial intelligence.

Shooting star—A popular name for a meteor.

Silicates—Rock-forming minerals that contain silicon and oxygen atoms, plus one or more other common types of atoms.

Sinter—An area rich in minerals such as carbonates, deposited by outflows from thermal springs, which produced the water that carried the minerals to the point of deposit.

Skepticism—A questioning or doubting state of mind, which lies at the root of scientific inquiry into the cosmos.

Solar nebula—The protoplanetary disk that surrounded the protosun.

Solar system—The sun plus the objects that orbit the sun, including nine planets, their satellites, asteroids, meteoroids, comets, and interplanetary dust.

Solvent—A liquid capable of dissolving another substance; a liquid in which molecules can float and interact.

Species—A particular type of organism. The members of a species possess similar anatomical characteristics and can interbreed.

Glossary

Spectrometer—An instrument for careful observation and measurement of a spectrum.

Spectrum—The distribution of photons by frequency or wavelength, often shown as a graph of the number of photons with each particular frequency or wavelength.

Star—A mass of gas held together by self-gravitation, in whose center nuclear-fusion reactions produce kinetic energy that heats the entire star, causing its surface to glow.

Stromatolite—Layered rocks made from colonies of bacteria that live at the boundaries between water and rock sediments.

Sublimation—The transition from solid to gas without a passage through the liquid state.

Sulfide—A mineral made of molecules in which negative sulfur ions have bonded to one or more positive metallic ions.

Temperature—The measure of the average kinetic energy of random motion within a group of particles. On the absolute or Kelvin temperature scale, the temperature is directly proportional to the average kinetic energy per particle.

Thermophile—An organism that thrives at high temperatures.

Terraforming—Changing a planet, or a planet's large satellite, so that it more closely resembles Earth.

Virus—A complex of nucleic acids and protein molecules that can reproduce itself only within a "host" cell of another organism.

***Voyager* spacecraft**—The NASA spacecraft, named *Voyager 1* and *Voyager 2*, which were launched from Earth in 1978 and passed by Jupiter and Saturn a few years later. *Voyager 2* proceeded to pass by Uranus in 1986 and Neptune in 1989.

X rays—Photons with frequencies greater than those of ultraviolet but less than those of gamma rays.

FURTHER READING

Angel, Roger, and Woolf, Neville. "Searching for Life on Other Planets." *Scientific American*, April 1996.

Davies, Paul. *Are We Alone?* New York: Basic Books, 1995.

DeDuve, Christian. *Vital Dust: Life as a Cosmic Imperative.* New York: Basic Books, 1994.

Goldsmith, Donald. *Worlds Unnumbered: The Search for Extrasolar Planets.* Mill Valley, CA: University Science Books, 1997.

Goldsmith, Donald (ed.). *The Quest for Extraterrestrial Life.* Mill Valley, CA: University Science Books, 1980.

Goldsmith, Donald, and Owen, Tobias. *The Search for Life in the Universe (2d ed.).* Reading, MA: Addison-Wesley, 1992.

Kargel, Jeffrey, and Strom, Robert. "Global Climatic Change on Mars." *Scientific American*, November 1996.

Klass, Philip. *UFO Abductions: A Dangerous Game.* Buffalo: Prometheus Books, 1989.

"Life in the Universe." Special issue of *Scientific American*, September 1994.

McKay, David, et al. "Search for Past Life for Mars: Possible Relic Biogenic Activity in Martian Meteorite ALH 84001." *Science 273*, 924, 16 August 1996.

Morrison, David, and Owen, Tobias. *The Planetary System (2d ed.).* Reading, MA: Addison-Wesley, 1992.

Sagan, Carl. *The Demon-Haunted World: Science as a Candle in the Dark.* New York: Random House, 1996.

Sagan, Carl. *Pale Blue Dot.* New York: Random House, 1994.

INDEX

Index

Baikonur, 206
Baross, John, 141
Benzene, 88, 89
Bradbury, Ray, 214, 215
de la Brosse, Olivier, 234, 236
Bruno, Giordano, 235
Buseck, Peter, 85

C

Calcium, 13, 49, 104, 105
California Polytechnic State
University, 79
California, University of, 5, 45,
68, 76, 82
Carbon, 13, 49, 51, 65, 88, 89,
104, 146, 147, 155, 186,
188
isotopes of, 50–52, 54, 93
Carbonaceous chondrite, 60,
105, 148
Carbonate, 13, 14, 20, 49, 65,
76, 81, 90, 104, 105, 109,
116, 163
Carbon dating, 52
Carbon dioxide, 13, 55, 104,
134–139, 144, 150, 159,
163–172, 175, 182–188,
219–221
Carbon monoxide, 55, 150, 169,
186
Case Western Reserve University,
115
Cell, 108, 112, 122, 137, 179,
189
Centimeter, 31
Challenger, 135
Chassigny, 57

Chicago Center for Religion and
Science, 235
Chicago, University of, 7, 49, 56,
145
Chillier, Xavier, 85
Chladni, Ernst, 3,
Chlorofluorocarbons (CFCs), 220
Chromium, 155
Claremont School of Theology,
233, 235
Clayton, Robert, 49, 56
Clemett, Simon, 82–85
Clinton, Bill, 3, 21, 217
Cobb, John, 233, 235
Cold fusion, 6–8, 91, 232
Colorado, University of, 83, 84,
212
Comet, 39, 139, 155, 161, 168
Contamination, 17, 19, 75, 86,
218
Cooper, A.S., 90
Cosmic rays, 32, 153
Cosmic seeding, 121, 151
Cosmic zoo, 154
Coxiella, 112
Crary, Albert, 46
Crick, Francis, 154, 221
Cross-linking molecules,
122–124, 126, 128, 180
Cryptoendoliths, 181, 182
Cytosine, 123

D

Dante Alighieri, 182
Dao Vallis, 201, 202
Darwin, Charles, 126, 142
Daughter nucleus, 52
Davis, Jefferson, 228

Index

Index

Index

Index

Index

Owen, Tobias, 9, 25, 26, 56, 102, 109, 198–200
Oxidation, 151
Oxygen, 13, 30, 49, 54, 56, 64, 78–81, 104, 134, 135, 138, 146–151, 155, 163, 168, 187, 188, 219–221

P

Pace, Norman, 142, 144
PAHs (polycyclic aromatic hydrocarbons), 13, 14, 20, 88–90, 93, 99, 101, 114
Panspermia, 151–155, 180, 239
Parent nucleus, 52
Pathfinder spacecraft, *see Mars Pathfinder*
Peary, Robert, 38
Permafrost, 166, 176
Perseid meteor shower, 37
Phenylbenzene, 89
Phobos, 23
Phobos spacecraft, 197, 198
Pillinger, Colin, 93
Planet, 21, 237
Planet-B spacecraft, 208
Planetesimal, 161–163, 168
Plate tectonics, 9
Polycyclic aromatic hydrocarbons, *see* PAHs
Pons, Stanley, 6
Potassium, 52, 53, 76, 153
Prokaryote, 130–132
Protein, 109, 126, 128, 146, 147
Proton, 50
Pyrolitic release experiment, 185, 188
Pyrrhotite, 78, 79

Q

Quadrantid meteor shower, 37

R

Radioactive decay, 33, 51–54, 75, 153, 186, 188
Radiometric dating, 33
Reckling Moraine, 42
Replication, 123, 124, 128, 179
Ribosome, 109
Rice University, 68
RNA (ribonucleic acid), 127–132, 148
Romanek, Christopher, 66, 67, 71, 80, 82, 115
Ross Desert, 181
Rubidium, 53, 76, 153
Russell, Bertrand, 222

S

Sagan, Carl, 6, 102
Salk Institute, 154
Saturn, 20, 22, 162
Schiaparelli, Giovanni, 23
Schopf, J. William, 5, 6, 82, 101, 102, 108–110, 117, 137, 189
Score, Roberta, 44–48, 50, 58, 63
Sedimentary rocks, 135, 174, 210
Shakespeare, William, 172
Shergotty, 57
Shooting star, 34, 37

Index

Xenon, 55, 56

Z